Agency and Cognitive Development

Oxford Series in Cognitive Development

Series Editors
Paul Bloom and Susan A. Gelman

What Babies Know
Volume I: Core Knowledge and Composition
Elizabeth S. Spelke

Core Knowledge and Conceptual Change
Edited by David Barner and Andrew Scott Baron

Causal Learning
Psychology, Philosophy, and Computation
Edited by Alison Gopnik and Laura Schulz

Making Minds
How Theory of Mind Develops
Henry M. Wellman

The Origins of Concepts
Susan Carey

The Foundations of Mind
Origins of Conceptual Thought
Jean Matter Mandler

The Essential Child
Origins of Essentialism in Everyday Thought
Susan A. Gelman

Agency and Cognitive Development

MICHAEL TOMASELLO

OXFORD
UNIVERSITY PRESS

Great Clarendon Street, Oxford, OX2 6DP,
United Kingdom

Oxford University Press is a department of the University of Oxford.
It furthers the University's objective of excellence in research, scholarship,
and education by publishing worldwide. Oxford is a registered trade mark of
Oxford University Press in the UK and in certain other countries

© Michael Tomasello 2024

The moral rights of the author have been asserted

All rights reserved. No part of this publication may be reproduced, stored in
a retrieval system, or transmitted, in any form or by any means, without the
prior permission in writing of Oxford University Press, or as expressly permitted
by law, by licence or under terms agreed with the appropriate reprographics
rights organization. Enquiries concerning reproduction outside the scope of the
above should be sent to the Rights Department, Oxford University Press, at the
address above

You must not circulate this work in any other form
and you must impose this same condition on any acquirer

Published in the United States of America by Oxford University Press
198 Madison Avenue, New York, NY 10016, United States of America

British Library Cataloguing in Publication Data

Data available

Library of Congress Control Number: 2024941635

ISBN 9780198896579

DOI: 10.1093/9780191998294.001.0001

Printed and bound in the UK by
Clays Ltd, Elcograf S.p.A.

Oxford University Press makes no representation, express or implied,
that the drug dosages in this book are correct. Readers must therefore always check
the product information and clinical procedures with the most up-to-date
published product information and data sheets provided by the manufacturers
and the most recent codes of conduct and safety regulations. The authors and
the publishers do not accept responsibility or legal liability for any errors in the
text or for the misuse or misapplication of material in this work. Except where
otherwise stated, drug dosages and recommendations are for the non-pregnant
adult who is not breast-feeding

For Rita and Chiara

In the beginning was the deed.
                              Johann Wolfgang von Goethe

Acknowledgments

Writing a book goes better when there are good conversational partners. For this book, I would like to thank Tamar Kushnir for many (walking through Duke Forest) conversations about key issues in cognitive-developmental psychology. I thank Dan Sperber and Pierre Jacob for many (walking through Luxembourg Gardens) conversations about the larger philosophical issues involved. And I thank Claudia Roebers for an enlightening conversation about metacognition (at a lovely café in Bern) at a critical early juncture. For commentary on the manuscript itself, I thank Henrike Moll and Hannes Rakoczy for helpful feedback on several key chapters and Jan Engelmann, Wouter Wolf, and an anonymous reviewer invited by Oxford University Press for helpful feedback on the manuscript as a whole. Institutionally, I thank Duke University for financially supporting a sabbatical year, and the Institut Jean Nicod in Paris for providing a rich intellectual environment for the second half of that year. And as always, I owe my deepest gratitude to my wife, Rita Svetlova, for her insightful thoughts and critical comments—as well as for her support and patience—throughout.

Contents

List of Figures   xi

1. Not By Learning Alone   1
2. Agency and Cognition   11

I EARLY INFANCY

3. Goal-Directed Agency and Iconic Representations   27

II TODDLERHOOD

4. Intentional Agency and Imaginative Representations   47
5. Joint Agency and Perspectival Representations   71

III EARLY CHILDHOOD

6. Metacognitive Agency and Multi-Perspectival Representations   103
7. Collective Agency and Objective/Normative Representations   127

IV MOVING FORWARD

8. An Agency-Based Model of Human Cognitive Development   155
9. The Child-as-Scientist Revisited   187

References   191
Index   207

Figures

2.1. The three types of psychological agency in the evolutionary line leading to humans, ordered bottom to top, as proposed by Tomasello (2022a).   16

2.2. The two basic types of human shared agency: (a) joint agency/intentionality among individuals and (b) collective agency/intentionality within a group.   19

3.1. Graphic depiction of young infants' image-schematic (iconic) representations of some basic kinds of events and object relations in their physical and social worlds (adapted from Mandler, 1992).   31

5.1. The perspectival representations underlying one child's use of six related verbs at around the second birthday. Each verb comprises two or three sequential "moments of attention" (one per rectangular box). The larger person represents the adult, the smaller person represents the child, and the oval represents an object. The arrow represents the causal force and direction of object movement. For example, *give* is the situation in which the adult causes an object to go from herself to the child. The shaded box indicates that the designated word profiles only the end state of the process.   88

5.2. From bottom: basic capacities for joint agency with its roles and perspectives (light gray box). These are foundational for cooperative-referential communication with the pointing gesture with their use of roles and perspectives in communicative intentions (medium gray boxes). These are in turn foundational for children's earliest perspectival symbols and role-based schemas in a conventional language (darker boxes), as well as more complex constructions that mark grammatical roles and discourse perspectives with second-order conventions (black boxes).   96

6.1. Graphic depiction of the current model of executive processes, including what is regulated, how it is regulated, and where, in what workspace, it is regulated (see Tomasello, in press).   107

6.2. Graphic depictions of three multi-perspectival concepts from early childhood. Arrows depict different perspectives; the box indicates a concept requiring coordination of two perspectives.   124

7.1. Highly oversimplified graphic depictions of three objective/normative concepts from early childhood. Arrows depict different perspectives or preferences or commitments; the box indicates objective/normative ideal. Persp = perspective; Pref = preference; Commit = commitment.   151

xii FIGURES

8.1. Summary of the three proposed steps of agentive organization in human ontogeny, along with their most distinctive cognitive capacities. First is the goal-directed agency of infants (bottom); second is the intentional agency of toddlers (middle); and third is the rational/metacognitive agency of preschool youngsters (top). 'JI' designates processes derived from the joint intentionality of toddlers, and 'CI' designates processes derived from the collective intentionality of preschoolers. 157

8.2. Summary of the hypothesized evolutionary sources of key components of human agentive and cognitive development (S = O indicates the notion of self-other equivalence). The leftmost diagonal labels indicate the most general level of agentive and cognitive organization. 171

8.3. Summary of different kinds of individual and social learning at different periods of human development (made possible by different forms of agentive organization). 173

8.4. Summary of factors determining the age of emergence of any given knowledge or concept. 181

1
Not By Learning Alone

Children of different ages live in different worlds. Consider a family trip to a zoo. The six-month-old infant perceives a new world of physical objects, animals, and people moving in 3D space. The two-year-old brother, in addition, imagines all kinds of hypothetical possibilities in the zoo based on an underlying causality of physical events and intentionality of agentive actions. He imagines his mother's perspective on a scene and coordinates with it in joint attention. The four-year-old sister, in addition, understands the zoo animals as falling into a hierarchy of species and genera and discerns what they believe about the world, even if it is not true. She feels the obligatory pressure of things that zoo visitors must and must not do based on the normative force of local rules and cultural norms.

Our question is how, over age, these three children have come to live in such different experiential worlds. A key part of the answer is of course learning: as children learn more and more about the world, they experience it in new ways. But it is difficult to see how an accumulation of learning could lead to systematic shifts in the *kinds* of knowledge and concepts with which children of different ages operate, including knowledge not only of objects and events in the perceptual world but also of causal possibilities, social perspectives, and normative necessities latent in that world. Given that children's learning cannot create new dimensions of experience *ex nihilo*, perhaps the answer to our question lies in maturational changes in the underlying architecture of children's cognition, that is, in the formats of cognitive representation, types of rational inference, and modes of self-reflection characteristic of children of different ages. Plausibly, it is changes in this architecture that structure most directly what children can experience and so learn.

In what follows I explore the hypothesis that there are indeed qualitative changes in how children of different ages represent their knowledge, make inferences from their knowledge, and reflect on and modify their knowledge. These changes are not independent of one another but work together at any given age in a coherent psychological organization originally evolved in one of humans' ancient animal ancestors for agentive decision-making and action. My attempt in the main body of this work is to specify the nature of these agentive architectures and how, as they unfold over time, they re-frame children's experience of the world. Cognitive development is more than an accumulation of learning—or so I will argue.

1.1. Brief Historical Background

The most influential theorist of children's cognitive development for most of the twentieth century, in both Europe and America, was Jean Piaget. He had a kind of answer to our question: children's cognitive development progresses through four distinct stages, from the sensory-motor cognition of infants through the ever more complex mental operations of older children and adults. But Piaget's formal description of these stages using mathematical group theory was limited and limiting in a number of ways, and it turned out that many of children's cognitive competencies did not fit neatly into his scheme. In addition, a growing appreciation for Vygotskian theory and research suggested that Piaget underestimated the social and cultural dimensions of children's cognitive development.

From the ashes of Piaget's theory, a new theory of human cognitive development has emerged. The originating idea was that children develop cognitively by operating like scientists. Thus, Carey (1985) detailed how children's cognitive development parallels scientific progress, and Gopnik and Meltzoff (1997) emphasized parallels in the ways that scientists and young children formulate and test hypotheses about how their concepts fit the world. Combined with the insight from cognitive psychology that concepts are always embedded in larger theories, this approach came to be known as the theory theory of cognitive development, of which there were several variants. What has recently brought unity to these variants are quasi-formal models of Bayesian learning and inference. Similar to scientists, children begin their cognitive journeys with some built-in "prior" expectations about what should be observed given what is already known—starting in early infancy with built-in core cognition of one type or another (Spelke, 2022)—and they go on to use these expectations to actively formulate and test hypotheses about how the world works in particular domains. The process is one of continuous and cumulative learning, with no qualitatively distinct across-the-board stages. Versions of this general approach are proposed by Carey (2009), Gopnik and Wellman (2012), Tenenbaum et al. (2011), and Xu (2019). Although there are many debates about particular issues, this general approach may simply be called the modern theory of cognitive development, and it has made great progress in establishing how young children learn about the world in many and various domains of knowledge.

What the modern theory has done less well, however, is to answer the basic question of why children develop the kinds of knowledge and concepts they do at the ages they do. By focusing almost exclusively on learning as the engine of cognitive development (given some built-in cognitive priors in infancy), the modern theory implies—or at least does not contradict—the proposition that with sufficient experience a child could learn anything at any age. That is to say, if we could develop a training regime that compressed children's real-world experience over many years into a several-week course, even a 1-year-old could master concepts such as *natural number*, *fairness*, or *game rule* (e.g., in soccer or chess). It is likely that many or most modern theorists would reject this proposition—1-year-olds are almost certainly not ready to learn these things—but it is not clear that the theory has the conceptual resources to reject it on principled grounds, much less to give a positive account of why children learn what they do when they do. One possibility might be to propose that some domain-neutral skill such as executive working memory matures during early childhood in ways that facilitate children's ability to learn more complex and abstract content over developmental time. But, to my knowledge, there exist no systematic proposals of this type, and moreover, it is difficult to see how expanded working memory could suddenly make it possible for young children to learn concepts involving qualitatively new conceptual dimensions, such as the normative dimension inherent in concepts such as *fairness* and *game rule*.

An alternative proposal is that what determines children's readiness to acquire certain knowledge and concepts at certain ages is their overall psychological organization, or architecture, and this develops with age. The reason that this proposal is not part of modern theory is likely because it smacks of Piagetian stages, which were so decisively rejected decades ago, and because modern theory has investigated domain-specific knowledge with such great success (e.g., starting with Hirschfeld & Gelman, 1994). But the fact that Piagetian stages are untenable does not mean that there are no qualitative, across-the-board transformations in cognitive development based on changes in overall psychological organization. Although some modern theorists have proposed qualitative shifts in knowledge and/or cognitive representations within knowledge domains (e.g., Carey, 2009; Gopnik & Wellman, 2012), these are different from across-the-board transformations or reorganizations. The only proposal in modern theory for an across-the-board shift is the claim that linguistic symbols provide a new format for cognitive representation that transition young children from the perception-based iconic representations of infancy to more adult-like conceptual representations (e.g., Carey, 2009; Xu, 2019). But whereas it is true that linguistic symbols facilitate the acquisition of concepts, they do not create qualitatively new conceptualizations out of nothing.

Thus, while these modern proposals for qualitative change are significant and in the direction at which I am aiming, they do not go far enough. I believe that to capture the across-the-board qualitative differences that characterize children's cognition at different ages we must step back and take a broader view of the overall psychological organization within which humans operate. Arguably the most comprehensive standpoint from which to take this view is evolution.

1.2. Agentive Organization

From an evolutionary point of view, cognition and learning exist for one and only one reason: to facilitate effective action. A species' cognitive and learning capacities are therefore organized around the individual's behavioral decision-making: its capacity to decide what to do, what we may call its agency. The computational model for agentive decision-making is not the digital computer, which stores information passively and only acts in response to external instructions. Rather, the computational model for agency is machines that behave autonomously and intelligently to actively pursue and maintain goals, the prototype being thermostats, self-driving cars, and other cybernetic control systems.

Autonomous and intelligent action requires the agent to perceive and comprehend goal-relevant aspects of the environment—as well as its own actions and their effects—in order to make informed and effective behavioral decisions fine-tuned to dynamically changing situations (see, e.g., Miller, Galanter, & Pribram, 1960; Gershman, Horvitz, & Tenenbaum, 2015; and a variety of other agent-based models). In such agent-based models (as in evolution) the function of learning is always and only to facilitate future behavioral decision-making. Overall, agentive organization provides the broadest vantage point for viewing cognitive development because it integrates all of the major domain-neutral psychological processes into one coherent adaptive system: (i) perception/attention/conceptualization as inputs to the decision-making process; (ii) action and learning as outputs (or results) of the decision-making process; and (iii) goal/intention/motivation/emotion as the relevance-providing direction for everything. In this view, developing children are not best characterized as Bayesian learners but rather, more generally, as Bayesian agents who learn in support of their agency.

A key component in much of the agentive decision-making of many organisms is an additional tier of executive monitoring and control. While the modern theory of human cognitive development recognizes executive function as important to cognition, it is typically conceptualized as a motley collection of separate processes (e.g., inhibition, working memory, cognitive flexibility, etc.). In contrast, Tomasello (2022a) argues that proactive executive processes have evolved to operate as components in an integrated tier of executive regulation that is itself a control system, whose function is to facilitate decision-making in situations of uncertainty. It does this by cognitively simulating and evaluating behavioral options—that is, by thinking and planning—to anticipate and pre-correct potential errors. Moreover, what is typically called metacognition may be seen as a kind of second-order tier of executive regulation that serves to facilitate the functioning of the executive tier. It does this by monitoring the organism's executive-tier thinking and planning and then intervening to improve them (e.g., by metacognitive belief revision). The historical scenario proposed for human evolution is that the earliest vertebrates operated as simple goal-directed agents with no proactive executive regulation, the earliest mammals operated as intentional agents with a single tier of proactive executive regulation, and the earliest great apes operated as rational agents with an additional, second-order tier of metacognitive regulation—with humans evolving on top of these some species-unique forms of shared agency involving unique forms of cooperative/normative regulation. Each of these forms of psychological agency is geared toward a particular type of agentive decision-making, requiring a particular format of cognitive representation, a particular type of rational inference, and particular capacities for self-regulation and learning.

My hypothesis is that these different forms of agency are substantially conserved in human cognitive ontogeny. At first glance, it is an improbable hypothesis because human infants and young children are so poor for so long at acting agentively on their own. But the claim is that they nevertheless operate with certain systems of agentive organization at certain ages—even if their competence for behavioral expression lags behind—and this structures and constrains their cognitive development and learning. The more specific hypothesis (Tomasello, 2024) is that young infants until nine months of age operate as simple goal-directed agents, making go/no-go decisions about whether to do something, with no skills of proactive executive regulation and with skills of learning focused on observed contingencies. With the emergence of a tier of proactive executive regulation at around nine months of age, toddlers begin operating as intentional agents—capable of thinking and planning before acting—to make more complex either/or decisions about what precisely to do, which spawns new forms of curiosity and exploratory learning structured by a means-ends causal analysis of both observed and imagined contingencies (which supports children's first real theories). And with the emergence of a second-order metacognitive tier of executive regulation at around three years of age, preschoolers begin operating as metacognitive agents who evaluate whether already-made decisions, or existing procedures for making decisions, are adequate to the task at hand, with learning now including belief revision (in the face of discrepant information) concerning generic or normative facts about the world.

As each new type of agentive organization emerges—in both phylogeny and ontogeny—the need arises for a new format of cognitive representation. That is because in different decision-making architectures, cognitive representations are doing different work. The current account thus posits: (i) perception-based iconic representations in early infancy for recognizing relevant situations (in order to make go/no-go decisions about whether to do something); (ii) imaginative and perspectival representations in toddlerhood for executively simulating and planning behavioral decisions proactively (in order to make either/or decisions about what precisely to do); and (iii) multi-perspectival and objective/normative metarepresentations in early childhood for comparatively evaluating beliefs and actions with respect to some ideal standard (in order to make decisions about what one *ought* to believe or do). These three formats of cognitive representation structure children's experience in ways that enable them, sequentially across age, to conceptualize things as existing in one of three fundamental modalities (first characterized by Kant, 1781, and still used in modal logic today): first as **actualities** (via the perception-based iconic representations of early infancy), then as **possibilities** (via the imaginative/perspectival representations of toddlerhood), and finally as normative **necessities** (via the objective/normative metarepresentations of early childhood). These qualitative shifts of agentive organization and representation do not concern specific cognitive content or knowledge—they do not concern all the rich cognitive content that children are learning all day every day—but only the domain-neutral agentive architectures that structure experience and learning.

This agency-based model also supports mechanisms of cognitive-developmental change that go beyond classical forms of learning. That is, Xu (2019) and others have recently begun exploring a set of cognitive processes dubbed "constructive thinking," in part to explain where Bayesian hypotheses come from in the first place. The idea is that as children are thinking about what to do or how something works, they are constructing new conceptualizations via such actively agentive processes as formulating hypotheses, articulating explanations, creating analogies across domains, and coordinating different perspectives. These processes do not originate in environmental input, but rather they are endogenous processes of self-regulation involving the creative coordination and/or reorganization of already existing knowledge and concepts: the child conceptualizes something in a new way by reflecting on and reordering what, in some sense, she already knows. This may even occur "offline" (i.e., outside of relevant goal-directed action) as the child seeks to eliminate discrepancies or redundancies in her existing knowledge representations. Karmiloff-Smith (1992) calls the outcome of such processes "representational redescription" or more simply, "re-representation," and she emphasizes that the process is not environmentally driven. In my view, processes of constructive thinking and re-representation are needed—along with learning, of course—to explain such things as preschoolers' transition from an approximate number sense to the concept of *natural number* (coordinating and integrating cardinal and ordinal perspectives), their transition from an understanding of the knowledge states of others to the concept of *belief* (coordinating and integrating different perspectives on the same situation), and their transition from an appreciation for the preferences of others to a normative concept of *fairness* (coordinating and integrating competing preferences among coequal individuals). The problem with existing accounts of constructive thinking and re-representation is that they are not systematic, and they do not specify the different ways in which these self-regulatory processes operate at particular developmental periods. By integrating constructive thinking and re-representation into an agency-based model—that is, a developmental series of such models—the current account is both more systematic and more developmentally sensitive.

Finally, as alluded to earlier, beyond individual agency human children also develop the capacity to participate in shared or "we" agencies underlain by cognitive processes of shared intentionality (Tomasello, 2019). Although the modern theory of cognitive development recognizes a variety of processes of social cognition and social learning (e.g., Gweon, 2021), it does not appreciate sufficiently, in my view, the indispensable role of mental coordination and social self-regulation (within shared agencies) in children's cognitive development. The point is that within shared agencies children do not just read other minds; they coordinate with other minds by taking into account others' perspectives, often on their own perspective recursively. They also work with partners in shared agencies to self-regulate their actions and decisions cooperatively. Because modern cognitive-developmental theory mostly neglects processes of cooperative coordination and self-regulation, it is not equipped to explain the emergence in early toddlerhood of such skills as joint attention, socially recursive inferences, and cooperative/referential communication. Nor can modern theory explain the emergence in early childhood of normative concepts, a sense of obligation, and the provision of reasons and justifications for beliefs. In terms of cognitive representation, cooperative coordination and regulation within shared agencies empower toddlers to develop perspectival concepts such as *chase* vs. *flee*, *over* vs. *under*, and *I* vs. *you* (requiring coordination of attentional perspectives), and they lead preschool youngsters to develop objective/normative concepts such as *true* vs. *false* beliefs, *right* vs. *wrong* actions, and *fair* vs. *unfair* outcomes (requiring coordination and regulation of conceptual perspectives and/or normative values).

Unlike human evolution, during which shared agency emerged only after all three forms of individual agency had evolved, in ontogeny joint agency develops early, at around 9–12 months of age, in conjunction with intentional agency and executive regulation, and this is then followed by group-minded collective agency at around 3–4 years of age in conjunction with metacognitive agency and metacognitive regulation. Therefore, at the most general level of analysis, the proposal is that there are two qualitative shifts in early cognitive development: (i) at around 9–12 months of age toddlers become capable of executively regulating their own actions and attention and also coordinating their action and attention with others in joint agencies; and then (ii) at around 3–4 years of age preschool youngsters become capable of metacognitively regulating their own executive-tier cognitive processes and also coordinating executive-tier cognitive processes in group-minded collective agencies. The three developmental periods created by these two organizational shifts— conventionally called early infancy, toddlerhood, and early childhood—occur in a generally similar manner in all cultural environments, each inaugurating a new way of operating that supports new capacities for decision-making, cognitive representation, and learning.

1.3. A Three-Part Proposal

The "core cognition + Bayesian learning" formula characteristic of modern cognitive-developmental theory constitutes a great advance in our understanding of children's cognitive development. My goal is not to replace this successful formula but rather to situate it within the wider context of children's development as functioning agents who use their learning to make effective behavioral decisions. Toward that end, I offer here a three-part proposal:

(i) human cognitive development unfolds in a sequence of three organizational architectures—originally evolved in ancient animal ancestors for different types of agentive decision making and action—and these structure children's experience and learning in different ways at different ages;
(ii) within these agentive architectures and their associated formats of cognitive representation and types of rational inference, children develop not only by learning from the environment but also by creating for themselves new hypotheses and theories—via self-regulative processes of constructive thinking and re-representation—that serve to reorganize their knowledge and concepts;
(iii) participation in uniquely human joint and collective agencies engages children in species-unique processes of mental coordination and social self-regulation that require novel formats of perspectival and objective/normative representation.

In articulating and defending this agency-based account, I do not attempt to review all of the rich and diverse literature of cognitive-developmental psychology, as that would require many books. Rather, after some evolutionary preliminaries, my attempt is only to look at those processes of human cognitive development that most clearly reflect its organizational architecture at particular developmental periods and to bring coherence to those within a single theoretical framework. This account will constitute my answer to the question of why children develop just the kinds of knowledge and concepts they do at just the ages they do.

2
Agency and Cognition

Animals are built for action. This is reflected in both their body morphology and their behavioral skills: lizards are built to capture insects, squirrels are built to cache nuts, and chimpanzees are built to use tools. It is also reflected in their supporting perceptual and cognitive abilities: lizards are built to recognize insects and track their actions, squirrels are built to identify nuts and find caching locations for them, and chimpanzees are built to understand the basic causal relations that make their tools effective. Animals perceive and understand the world in just the ways needed to support their adaptive actions.

These various behavioral and cognitive adaptations may be organized in different ways, depending on the nature and variability of the relevant ecological challenge. Thus, when a species faces an urgent and recurrent ecological challenge, natural selection tends to hardwire specific perception–action connections, creating reflexes and other stimulus-driven actions (think swallowing and breathing). But when the environment is more unpredictable, natural selection crafts an agentive architecture that empowers the individual to operate in a more flexible manner by cognitively representing desired goal-states, making perceptually and cognitively informed behavioral decisions toward those goal-states, and self-regulating the process throughout. Whereas stimulus-driven, reflexive actions are organized as linear stimulus-response chains, goal-directed actions are organized circularly in the manner of a control system: the organism pursues a goal by acting on the perceived situation and then monitors and evaluates the result, which then leads to goal-directed adjustments in a continuous cycle of feedback control.

Organisms that are built with control system behavioral organization—in at least some of their major activities—are what we may call psychological agents, who possess the flexible perceptual and cognitive capacities needed to assess situations and decide how to best pursue their goals. This organization may take several forms. Tomasello (2022a) proposes an evolutionary history of psychological agency in species leading to modern humans, comprising a graded series of architectures built on control system principles. To understand the ontogeny of human agency, and so the ontogeny of human cognition,

Agency and Cognitive Development. Michael Tomasello, Oxford University Press. © Michael Tomasello 2024.
DOI: 10.1093/9780191998294.003.0002

we must understand these architectures and how they structure cognitive representations, processes, and skills—since, by hypothesis, they are evolutionarily conserved in human ontogeny.

2.1. Agentive Organization and Decision-Making

Despite the dizzying array of particular behavioral and cognitive adaptations in the animal kingdom, the agentive organization of action and decision-making comes in only a few basic forms. The organizational backbone of all forms of psychological agency is control system architecture.

2.1.1. Control System Organization and Architecture

Modern digital computers are good at what they do, which is to store and retrieve information upon command and perform operations or computations on that information as instructed. But whenever humans want to build a machine that acts autonomously, like a living creature, we turn to cybernetic mechanisms organized as feedback control systems. If we want a machine to keep a constant temperature in a room, we must design it not only with a furnace and air-conditioner but also with a thermostat that can sense the temperature in the room (perception) and compare it to a human pre-set value (goal) and then operate the furnace or air-conditioner (action) as appropriate. Of course, we have electric fans that blow cool air and space heaters and that blow warm air, but they do not operate intelligently and autonomously, since they require a human to turn them on or off as she judges what is needed (i.e., the human acts as thermostat). Control systems organization was first discovered by the early cyberneticians Norbert Wiener and Ross Ashby and was made popular in psychology by Miller et al. (1960). Almost all modern agent-based models (e.g., Gershman et al., 2015) assume this basic control system architecture in which goal (or reference value), perception, and action operate as integrated components in producing intelligent and autonomous behavior.

For simple machines, information for the decision-making mechanism comes from some kind of simple sensing device like a thermometer. But for agentive organisms, things are more complicated. Agents are built to sense or perceive many things in the environment that could potentially be relevant for their actions in some context. But when it comes time to make a particular behavioral decision in a particular context, the organism *attends* to only some things, namely, those situations that are *relevant* for its behavioral decision-making in that they present opportunities or obstacles for goal realization. Attention is thus not simple perception, and, indeed, depending on its goal the organism may attend to multiple relevant situations in a single perceptual image. For example, a squirrel may perceive a nut in a tree and, being hungry and preparing for action, may attend to the facts that: (i) the nut is ripe for eating; (ii) the nut is in an accessible location; but (iii) there is a competitor nearby. It is noteworthy that the goal-directed organism does not attend to objects or properties simpliciter but rather to goal-relevant *situations* as configural wholes: *the fact that* the nut is accessible.[1] Attending to whole situations thus constitutes a kind of quasi-propositional structuring of experience (Tomasello, 2014).

Situations in the environment that are relevant for an organism's (potential) goal-directed actions constitute its "experiential niche" (Tomasello, 2022b). Of course organisms may sometimes attend to objects, events, and properties outside of goal-directed decision-making, for example, when captured by so-called bottom-up attention (e.g., a sudden noise behind them or an enticing smell in front of them). But importantly, one could think of bottom-up attention as attention to situations that have been flagged by natural selection as especially relevant, even urgent, for the "goals" of survival and procreation, no matter one's immediate goal state. In addition, even more important for cognitive development, some organisms also attend to situations that are not of direct relevance to current goals when they are actively exploring the world out of curiosity. Nevertheless, even in such cases individuals will not be curious about and explore things that are completely irrelevant for their action capabilities: squirrels will not be curious about potential tools and lizards have no interest in nuts.

2.1.2. Three Types of Individual Agency

Tomasello (2022a) proposes that if we focus on the species forming an evolutionarily line to humans, there have been only three basic forms of individual agentive organization. Goal-directed agents are organized as control systems;

---

[1] It does this because goals themselves are internally represented as desired situations: the squirrel's goal is not the nut *per se* but rather having or eating the nut, and so it must identify the environmental situations that might promote or interfere with realization of this goal situation.

intentional agents are organized as control systems with an additional tier of executive supervision and control; and rational (or metacognitive) agents are organized as control systems with a further, second-order, metacognitive tier of executive supervision and control. Each type of agency is characterized by its own particular types of cognitive representations, inferential operations, and skills.

Goal-directed agency. The first terrestrial vertebrates were lizard-like creatures who faced the unpredictabilities of insects attempting to evade them. They thus operated as goal-directed agents: decision-makers organized as control systems to pursue their internally represented goal-states until they perceived that the environmental situation matched them. They made a go/no-go decision in each situation and learned from it. They were capable of a simple process of reactive inhibition (a global inhibition or "freeze" response), which enabled them to stop what they were doing, if need be, which put them in the position to make a new go/no-go decision in the new situation. They were restricted to this mode of decision-making because they had no executive tier of proactive executive control. These early vertebrates operated with perception-based iconic representations that enabled them to recognize goal-relevant objects, events, and situations, with a natural generalization to "similar" forms. These representations thus generated various kinds of inferences, as objects and events recognized as of the same kind were expected to have similar properties and do similar things. Anything that violated expectations elicited reactions of surprise, accompanied by redoubled attention.

Intentional agency. The first mammals were squirrel-like creatures who faced the unpredictabilities that come with life in a competitive social group. They thus operated as intentional agents: decision-makers able to mentally simulate possible action choices and likely environmental changes before acting. These simulations required a new form of imaginative representation that went beyond perception-based iconic representations by representing things and situations that were not currently the case. On the basis of these simulations, intentional agents made either/or decisions about which action to perform, and these then translated not directly into action but into an intention to act (which could then be more easily inhibited). Such simulations also enabled the construction of complex action plans with one action embedded within another (e.g., removing an obstacle on the way to a goal), the prototype of intentional action, as well as instrumental/causal learning of actions in which means and ends were distinguished. These processes of proactive decision-making were enabled by, and took place on, an evolutionarily new executive tier of functioning that was itself a control system, whose function it was to monitor the actions and attention of the organism and intervene as necessary to facilitate the decision-making process. These early mammals were thus able to engage in simple forms of thinking and planning (in an executive workspace), using imaginative representations, accompanied by more proactive types of inhibitory control (e.g., suppression of the unchosen behavioral option before acting).

Rational (metacognitive) agency. The first great apes were chimpanzee-like creatures who faced the unpredictabilities associated with intense contest competition for resources. They thus operated as rational agents: decision-makers able to reflect on decisions already made and assess their appropriateness given new information (belief revision). These processes of reflective control took place on an evolutionarily new, second-order, metacognitive tier of functioning that, again, was itself a control system whose function was to monitor the process of executive decision-making and intervene as necessary to facilitate the decision-making process even further. These early great apes could decide between two possible decisions metacognitively, for instance, when choosing between a decision based on less sufficient or more sufficient information, leading them in some cases to actively seek new information. Great apes' intense contest competition also led to the evolution of powerful skills of social learning that required individuals to compare their own intentional actions to those of others, which spawned the ability to attribute one's own manner of psychological functioning to other psychological agents. And skills of tool use gradually became more flexible and led to attributions of causality not only to one's own actions but also to external events.

In this evolutionary account, the three types of agentive architecture differ most fundamentally in the nature of the executive processes used to monitor and control decision-making (see center diagrams in the panels of Figure 2.1). Decision-making with one or the other of these architectures necessitates particular types of cognitive processes (see list on the right in Figure 2.1). Inspiration for this hierarchy of executive control systems comes from models in cognitive science that focus on processes of so-called cognitive control (see, e.g., the various papers in Egner, 2017), which provide more comprehensive and integrated accounts of agentive action than do models of so-called executive function, which tend to be tightly tied to neuropsychological testing. The current model thus situates and organizes executive and metacognitive processes of regulation in an agentive architecture with cognitive processes built for particular types of behavioral decision-making.

This hypothesized evolutionary trajectory proposes a natural buildup in complexity over evolutionary time, a common occurrence in biological

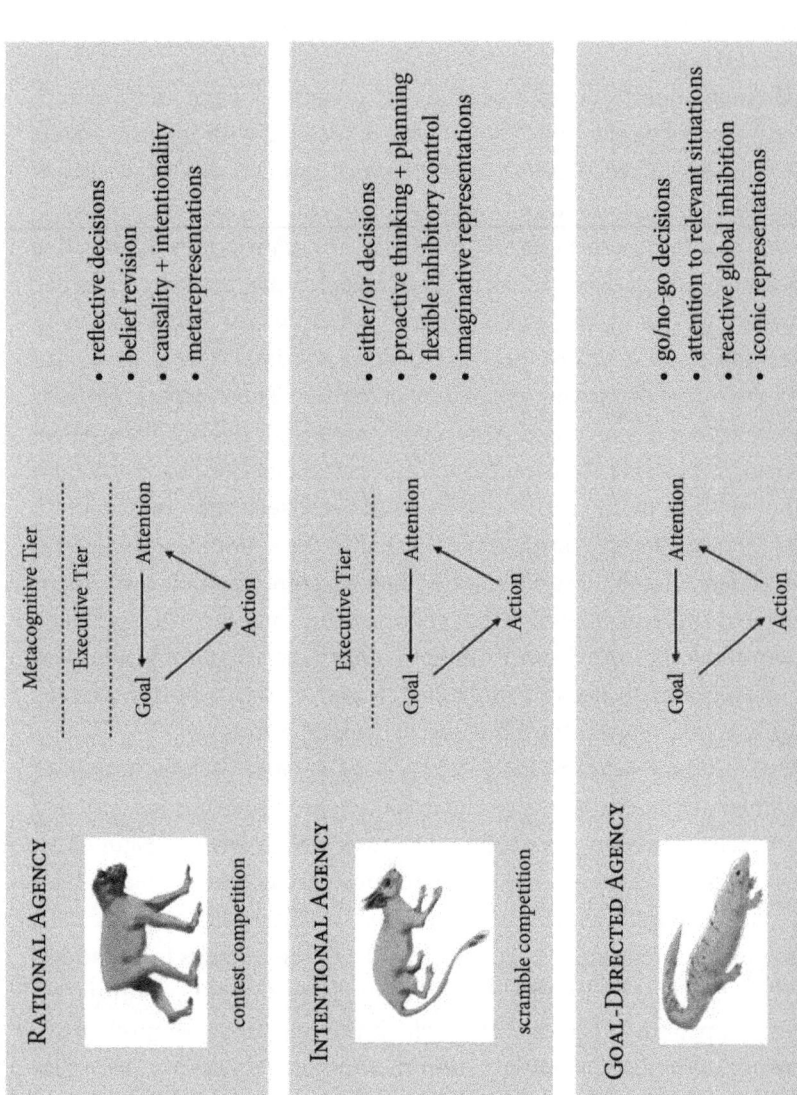

Figure 2.1. The three types of psychological agency in the evolutionary line leading to humans, ordered bottom to top, as proposed by Tomasello (2022a).

systems of all types in which subsequent forms build on already existing forms (Bonner, 1988). If this evolutionary account is broadly accurate, it might then be possible to observe in human ontogeny today these different forms of agency in the way the individual makes decisions, cognitively represents the world, and executively monitors and controls its actions (although there is nothing necessitating that human ontogeny unfolds in this way). And they might emerge in the same order since, as in the complexity account of evolution, the simpler forms are foundational for the more complex forms (e.g., a control system with two tiers of executive control presupposes a control system with one tier of executive control).

2.1.3. Two Types of Shared Agency

Humans' actions may be organized in all of these different ways, but in addition, humans have evolved two species-unique forms of agentive organization to accommodate the need to act cooperatively with others in special ways. These shared or "we" agencies require still other sets of cognitive representations, operations, and skills, which are responsible for most of the psychological characteristics that so clearly distinguish humans from other animal species (Tomasello, 2014, 2016).

Joint agency. Around one-half million years ago, early humans started a new chapter in the evolution of agency: they began to form with one another joint or "we" agencies, in which two individuals operated as a single *co*-operative agent acting toward joint goals, jointly attending to relevant situations (and thus creating joint knowledge or common ground) so as to make the best joint decisions. The joint agency thus constituted a dual-level cognitive structure of sharedness and individuality: "we" are working toward a joint goal with joint attention, but each of us has her own individual role and perspective in the process. To coordinate actions and attention, individuals needed to coordinate their mental states recursively, including in acts of cooperative communication (e.g., via a pointing gesture). This required them to form perspectival cognitive representations that reflected the different aspects of the situation to which each partner was attending. Because each partner could be tempted to opt out of the cooperation, the individuals also worked together to self-regulate the cooperation, for example, as one called out the other for uncooperative behavior or else they formed a joint commitment ahead of time to forestall uncooperative behavior. Operating in a joint agency required the individual to operate intentionally on an executive tier of functioning and to understand her partner as operating in that way as well: just as each regulated her own actions and attention executively, in a joint agency they coordinated and self-regulated each other's actions and attention executively as well. The underlying cognitive skills involved are thus termed joint intentionality and involve individuals making inferences and taking the perspective of one another recursively.

Collective agency. Then, one or two hundred thousand years ago, modern humans scaled up the cooperative process to the level of the entire cultural group, which acted as a kind of collective agent with collective goals and knowledge (cultural common ground) geared toward making the best collective decisions. This created a dual-level structure of a new type comprising both the group's collective "we" and the individual's role and perspective within it. To coordinate processes of cultural collaboration, which could involve interaction with in-group strangers, individuals agreed upon conventions and norms (and ultimately institutions) to which individuals had to conform. Key to the process of social coordination was conventional forms of communication, a.k.a., conventional languages, in which thoughts were expressed and examined publicly. Joint and collective decision-making in a conventional language—and the mental coordination that this entailed—led individuals to construct objective representations reflecting a kind of perspectival consensus (as well as other kinds of multi-perspectival representations). Because individuals could be tempted to opt out, the group worked together as a collective agent to self-regulate its individuals by enforcing social norms on all, including the self, creating a kind of we > me self-regulation of obligation to the group (and guilt for transgressions). Operating in a collective agency required individuals to operate rationally on a metacognitive tier of functioning and to understand others as operating in this way as well, regulated by the collective "we" and its normative ideals specifying what anyone who would be a member of the group *ought* to believe and do. The underlying cognitive skills involved are thus termed collective intentionality and involve individuals taking objective and normative perspectives on things, as often instantiated in cultural conventions and norms.

Figure 2.2 graphically depicts the two forms of shared agency. Contemporary human adults operate not only via both of these but also via the three forms of individual agency—in at least some of their activities on at least some level of analysis. So, for example, returning a book to a friend might involve the goal-directed activity of walking, embedded in the intentional activity of returning the book, embedded in the rational activity of deciding whether returning the book now is better than keeping it for a while longer. If I borrowed the book with my study partner, we would have to make a joint decision about whether

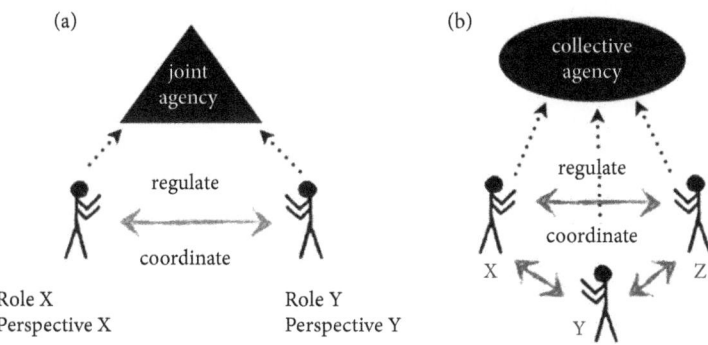

Figure 2.2. The two basic types of human shared agency: (a) joint agency/intentionality among individuals and (b) collective agency/intentionality within a group.

returning the book now is better than keeping it for a while longer, and we might mentally coordinate by communicating and justifying our opinions or even making a joint commitment to return the book tomorrow. And we might do all of this in the context of a sense of obligation to our friend and a healthy respect for the norms of reciprocity in our culture. Human individuals acting as joint agents or collective agents basically cooperativize the processes of individual agentive decision-making, as they *co*-operate to make shared decisions to pursue shared goals relying on shared attention and knowledge, as well as to self-regulate the process of normativity.

2.1.4. Evidence for the Evolutionary Story

Tomasello (2022a) reviews a wealth of behavioral data on contemporary model species to justify these different forms of individual agentive organization in ancient species leading to humans: lizards as exemplars of the first land vertebrates acting as goal-directed agents; squirrels as exemplars of the first mammals acting as intentional agents; and chimpanzees as exemplars of the first great apes acting as rational/metacognitive agents. The two forms of shared agency are connected to two early hominin species: *Homo heidelbergensis* as exemplars of the first joint agents, and *Homo sapiens sapiens* as exemplars of the first collective agents. There is no doubt that these model species and hominin exemplars are imperfect representatives of the idealized forms of agency posited. The hypothesis is that in the actual evolutionary sequence there was a gradual transition from one form of agency to another in a perfectly normal process of evolution by means of natural selection.

2.2. Human Cognitive Ontogeny

Ontogeny does not have to recapitulate phylogeny, but it often does (Gould, 1977). This is because new species conserve the effective developmental pathways of their progenitors, with new adaptations often modifying or building on existing adaptations after they are ontogenetically expressed. The general rule is that the more basic and fundamental a developmental pathway, the more likely it is to be conserved across successor species since changes in these would be more disruptive to other important pathways. Thus, the human body inherits its bilateral symmetry from ancient Cambrian ancestors some 500 million years ago and its backbone from ancient vertebrate ancestors some 300 million years ago. Less basic and fundamental adaptations enter into ontogeny at different times depending on many factors.

2.2.1. Developmental Timing

The evolutionary hypothesis is that control system organization is the backbone, as it were, of vertebrate psychological agency. Mammalian psychological agency then built an executive tier of oversight and control on top of this, and great ape psychological agency then built a metacognitive tier of supervision and control on top of them both. Given this foundational psychological framework, other traits and capacities may shift in their developmental timing (within limits) across successor species; for example, humans begin self-produced locomotion at a later age than other great apes but follow the gaze direction of others at an earlier age (Tomasello, 2020a). And new traits in successor species—for example, habitual tool use as a novelty in great apes among mammals—may enter into the conserved ontogenetic sequence at any point, so long as all necessary prerequisites are in place and it does not disrupt other critical pathways (West-Eberhard, 2003).

The most distinctive characteristic of human ontogeny is how slowly it unfolds in comparison to that of other vertebrates, mammals, and primates. One consequence of this extended immaturity is that some temporal decouplings may occur in ontogeny that could never occur in evolution. For example, because human infants must be born into the world before they are overly large—for anatomical reasons independent of psychology—they are motorically incompetent for some time. Caregivers compensate for this by, essentially, acting as agents for them. But some of the perceptual and cognitive skills that evolved to support goal-directed action nevertheless emerge before they are needed for action in the life of the immature infant herself, presumably because they are needed to support the learning of things that will be important when she later becomes more behaviorally competent. Similarly, human toddlers display all kinds of species-unique cognitive capacities—in such things as social cognition, social learning, and communication—while they are still almost completely incompetent at any kind of serious decision-making in such things as obtaining food and avoiding dangers, and most of their complex social interactions take place only with competent adults, not with equally incompetent peers. Infants and toddlers thus possess cognitive abilities for exploring the world in ways that will be relevant for their later attempts to agentively exploit it (Gopnik, 2020). In general, particular periods in the ontogeny of a species may pair organizational architectures with behavioral and cognitive skills in ways that differ from the pairing in evolution.

With respect to individual agency, it appears that the ordering in human ontogeny does indeed reflect the ordering in phylogeny: first goal-directed agency, then intentional agency, then metacognitive agency (as I will attempt to show in upcoming chapters). With respect to shared agency, it appears again that the ordering in human ontogeny also reflects the ordering in phylogeny: joint agency precedes collective agency. But whereas both types of shared agency emerged in human evolution only after all three types of individual agency were already operative, in ontogeny something different happens. After some emotional preparations in early infancy, skills and motivations of joint agency/intentionality emerge along with intentional agency at 9–12 months of age—at least partly because they both require an executive tier of functioning. Then skills of collective agency/intentionality emerge along with metacognitive agency at around 3–4 years of age—at least partly because they both require a metacognitive tier of functioning. I flesh out in detail in the chapters that follow the interconnected development of individual and shared agencies in the context of the especially challenging social context within which human children live and learn, that is, involving multiple caregivers (unique among great apes; Hrdy, 2006). Tomasello and Gonzalez-Cabrera (2017) provide an evolutionary account in which adult skills of shared intentionality "migrated down" in ontogeny across generations during human evolution to meet these challenges.

An important question in all of this is why it is 9 months of age and 3 years of age for the key ontogenetic transitions in human agentive organization? My highly speculative proposal is that these ages are especially important in the exercise of children's independent agency. Nine months is the earliest age for skillful self-produced locomotion—skilled crawling and initial walking—which gives children increased independence from adults in much of their decision-making. And three years is the classical age of weaning in ancestral societies, which exposes toddlers to novel—and more complex—situations in which they must make independent decisions. After the age of weaning, young children become much more competent both at making their own individual decisions and at making cooperative decisions in shared agencies with co-equal peers.

2.2.2. The Developmental Hypothesis

The developmental hypothesis, then, is this. All of children's experience and knowledge (i.e., cognitive content in specific domains) function within a cognitive organization biologically adapted for one or another type of agentive decision-making and action, that is, within an agency-based, control system architecture comprising goals and intentions, perception and attention, cognitive representations and operations, decision-making and action, executive regulation and learning. As development proceeds, this organizational architecture changes from having zero to one to two tiers of executive supervision and control, each of which provides a new framework for cognitive representations and operations, as well as for learning and re-representation. Shared agencies operate within these same overall architectures, but with some additional features. In terms of ontogenetic timing, the hypothesis is that these different forms of agentive organization and their associated cognitive representations and operations naturally emerge along the following timeline:

- **Early Infancy.** Goal-directed agency with perception-based iconic representations emerges in early infancy and operates throughout the first nine months of life.
- **The 9–12-Month Transition.** Both intentional agency with imaginative representations and joint agency with perspectival representations emerge at 9 to 12 months of age and predominate in toddlerhood until about 3 or 4 years of age. These agencies are made possible by an executive tier on which the toddler self-regulates her attention and action via

proactive thinking and planning, and also coordinates her attention and action with others.
- **The 3-4-Year Transition.** Both metacognitive agency with multi-perspectival representations and collective agency with objective/normative representations emerge at 3 or 4 years of age and predominate in early childhood until about 6 years of age. These agencies are made possible by a metacognitive tier on which the preschool youngster self-regulates her executive-tier thinking and planning via the coordination of perspectives and also coordinates her thinking and planning with peers (both individually and in groups).

These emerging organizational frameworks structure how children experience and learn about the world and so structure the ontogeny of the many contentful domains that constitute human cognitive development. Of course, older ways of experiencing the world do not disappear when newer ones emerge but persist in ways that can sometimes create interesting cognitive conflicts. In terms of shared agencies, it is important that great apes operate with both an executive and a metacognitive tier of functioning but do not participate in shared agencies, a fact which helps to specify the precise contribution of shared agencies to human cognitive development.

The empirical argument for the age of emergence of these organizational frameworks is an "inference to best explanation" about the architecture needed to account for widely recognized transitions in human cognitive ontogeny (e.g., the age of emergence of tool use or joint attention or normative language). Importantly, the age at which some particular cognitive content or knowledge is mastered is not directly determined by the operative agentive framework—which only opens up possibilities for learning—but more directly by the individual's experience with the relevant cognitive content. Thus, children become capable of using tools at around their first birthday, but the age at which they master any particular tool varies enormously depending on, among other things, the cognitive complexity of the tool and the child's experience with it. In general, each new organizational framework brings into existence both new things to be learned and new ways of learning them.

I
EARLY INFANCY

Perception can have much more abstract content than one might initially think.

Susan Carey

NB: Infancy begins at birth, but it does not have a well-defined endpoint. In the current account, for reasons that will become clear, infancy ends at nine months of age. Because this is a few months younger than is conventional, I will often refer to "early infancy" and "young infants." But, reader beware, often I simply use the terms "infancy" and "infants."

3
Goal-Directed Agency and Iconic Representations

Human newborns are not very agentive; their actions are mostly limited to looking, sucking, and various reflexes. They survive because mothers and others act as agents for them to supply what they need. Even during this period of extreme helplessness, infants are understanding and learning much about the world.

Piaget (1952, 1954) misunderstood this situation. As a biologist, he knew that in evolution cognition is always in the service of action and agency. His mistake was the further conclusion that children's cognition develops today only as needed to support their agentive action (cf. Russell, 1996). But because adults act agentively for infants, this connection between cognition and action has, to some degree, been severed in human ontogeny (in a similar but more pronounced manner than in other mammals and primates). At the same time, life in a cultural group requires children to master many skills and much knowledge—orders of magnitude more than other mammals and primates—before they can function as fully independent agents. The outcome is that human infancy and early childhood have evolved to be periods for learning about the world free from the pressures of self-sufficiency.

But as research in ethology and comparative psychology has made clear, learning is never written on a blank slate. In Bayesian terms, learning always requires some "priors" as an initial foundation. In the past few decades, developmental research has identified many of those priors in human infancy in the form of built-in core cognition that structures infants' early experience of space, objects, quantities, and animate beings. This core cognition is expressed not in their (very limited) actions but in their attention to and expectations about the world.

Thus, although cognitive processes evolved in humans' animal ancestors to support agentive decision-making and action, in human ontogeny today these same processes develop earlier than the behavioral competencies they originally evolved to support. In this chapter, I follow the ontogenetic order and look first at infant cognition as expressed in attention and perceptual expectations in the first months of life, and then at later infant cognition as expressed in agentive decision-making and action. Importantly again, for reasons that will become clear later, I consider as infancy only the first nine months of life.

3.1. Before Agentive Actions

For the first few months of life, infants act on their own bodies (e.g., sucking their thumbs), but they do little to make things happen in the external world. However, ingenious research has shown that even young infants are organized psychologically as goal-directed agents. In experimentally contrived circumstances they are surprisingly skillful at producing and controlling external effects. For example, Kalnins and Bruner (1973; see also Rochat & Striano, 1999) showed one- to three-month-old infants a movie whose visual clarity they could control by sucking on a plastic nipple. They quickly learned to use their sucking agentively to gain control over the movie's clarity, suggesting that from soon after birth infants are cognitively capable of producing agentive actions in novel situations, even if their immature motor systems are incapable of producing them in most instances.

Newborns also come equipped to recognize and learn from certain kinds of experiences independent of their actions. Specifically, like many animal species they pay heightened attention when special dangers or opportunities appear in their immediate surroundings and also, more generally, when perceived situations violate their expectations—presumably because it is adaptive to notice when things have changed. This is the bottom-up attentional bias that human infancy researchers have exploited to such great effect, as we shall see in this section. In the two sections following this one we focus on how older infants also attend, top-down, to physical and social situations that they, as behaving agents, find relevant for their goal-directed actions. Both bottom-up and top-down attention are important in human cognitive development because it is not perception *per se* that leads to learning and knowledge but rather attention to situations that are relevant for the infant—as determined by either natural selection or by the acting infant's goals.

3.1.1. Attention to Objects and Events in Space

Exploiting infants' bottom-up attention, Spelke (1990) established that in their visual perception infants are sensitive to the Gestalt properties of objects such as cohesion, boundedness, and identity (i.e., they pay heightened attention to experimental objects that violate these principles). In addition, Baillargeon (1987) showed that infants as young as three months of age—before they are reliably manipulating objects manually (and are still failing behavior-based object permanence tasks)—can track an object's trajectory and infer its continued existence, even though at the moment they cannot perceive it beneath a board (i.e., they pay heightened attention if the board under which an object disappeared folds down flat). In addition, infants of this same age also know much about object quantities, if the number of objects involved is small (Xu & Spelke, 2000). This knowledge of objects, space, and quantities is common among many animal species—including reptiles and most other animals—which makes sense for animals who must forage for food in all kinds of novel locations (Wilkinson & Huber, 2012).

Infants also know much about events and some physical principles that govern those events. Again employing violation-of-expectation paradigms, researchers have shown, for example, that infants expect that unsupported objects will fall, that the movement of objects will be blocked by physical barriers, that one object may contain inside it another object, and other basic physical principles of object movement in space (see Baillargeon, 2008, for a review). Importantly, in many of these studies the key event happens behind an occluder, and then when the occluder is lowered and an impossible outcome is revealed, infants pay heightened attention. They thus make inferences about objects and events that they do not directly perceive. But they cannot translate that expectation into action. Thus, as a specific example, Spelke et al. (1992) found that three- and four-month-old infants expect an object to be supported by a shelf, but it is not until six months later that they can use an understanding of object support in behavioral tasks (Willatts, 1984, 1999). Infants' expectations about the world at these young ages support recognition, inference, and learning about objects and events, but not agentive decision-making and action.

3.1.2. Attention to Animate Beings

Almost from birth human infants are socially oriented. In looking-time experiments, infants of only a few weeks or months of age show a special attraction to faces, and they distinguish biological from inanimate motion in point-light displays. Behaviorally, they mimic several different adult facial movements, again suggesting a special attunement to animate beings, and they engage with other humans in various forms of emotional exchange and emotional attunement in ways that are simply not applicable to inanimate objects (see Rochat, 2004, for a review of these and other relevant studies). As in the case of objects and events in space, then, young infants come into the world with built-in expectations about other humans and some of their basic characteristics and actions.

Importantly, with both inanimate objects and animate beings, infants not only operate with the expectations that nature has built in, but they also develop expectations based on their own observations of the interactions of objects, people, and events in the world. Thus, in violation-of-expectation experiments, infants are given repeated experiences with a novel auditory or visual sequence so that they habituate to it and so build up expectations about it. Violation of these learned sequences also leads to heightened attention (behavior also shown by various bird and mammalian species; see Saffran & Kirkham, 2018, for a review). They also learn what to expect from particular human beings in their social environment (e.g., Mom versus Dad). Although neither built-in nor learned expectations create in infants an ability to act overtly or to make decisions behaviorally, they clearly reflect some kind of cognitive representations of the world. But what kind?

3.1.3. Iconic Representations for Recognition and Learning

The nature of infants' cognitive representations has been, and continues to be, a topic of serious controversy (see the papers in Rakison & Oakes, 2003, and the commentaries on Carey, 2011). Some researchers characterize them as perceptual because they represent infants' perceptual experiences of the world, whereas others consider them conceptual, highlighting the fact that they include nonperceptual information such as 'can provide support', or else their representational content is nonperceptual (e.g., quantity). Spelke et al. (2010, 2022) refer to infants' "core knowledge and concepts" and stress that they are informationally richer than concrete sensory images (although still less abstract than adult cognitive representations). Burge (2011) believes that the distinction between perceptual and conceptual is specious and that even things like "quantity" are, for infants, basically perceptual—if one has an appropriately rich theory of perception. Carey (2009, 2022) agrees and terms infants' representations "iconic" and characterizes them as "perceptual systems . . . with innate input analyzers." Xu (2019) attempts to have it both ways by calling infants' early cognitive representations "proto-conceptual primitives," whereas Mandler (2004) attempts to have it both ways by claiming that infants

Support Containment

Going Out Going In

Animate Motion

Figure 3.1. Graphic depiction of young infants' image-schematic (iconic) representations of some basic kinds of events and object relations in their physical and social worlds (adapted from Mandler, 1992).

have both perceptual and conceptual representations in the form of "image schemas" (see Figure 3.1).

The common thread in all these descriptions is that infants cognitively represent the world in a manner that is based on perception but is more abstract than any particular sensory content: the format of infant cognitive representations is perceptual but schematic or, more simply, iconic. Given this perception-based iconic format, it is misleading to characterize the content of infants' representations with words from adult language, as is done by some researchers. An alternative is to depict them graphically in iconic diagrams (whose elements constitute a theoretical metalanguage, with elements rigorously defined linguistically). Thus, Mandler (2004) represents infants' so-called image schemas as in Figure 3.1 depicting such things as object support, object containment, objects entering into and exiting from containers, and animate motion. Iconic diagrams such as these are appropriate for depicting infants' representations because infants' representations are themselves iconic (not linguistic). Many other vertebrates also cognitively represent their experience in this same format such that perceived deviations surprise them and attract redoubled attention (Spelke, 2022, often cites research with newborn chicks). Although the representations of different species of course have different content, it is nevertheless likely that the basic iconic *format* of human infants' cognitive representations has deep evolutionary roots going back at least to the earliest vertebrates.

Crane (2003) and other philosophers believe that iconic representations are problematic because the same image may represent a wide variety of potential referents (e.g., an apple, a fruit, an object). But this is only a problem if we consider a disembodied representation such as a picture (see Barsalou, 2008, for a perception-based view of adult cognitive representation). An iconic representation constructed by a cognitive organism is not a disembodied picture but rather a schematization of what it has attended to in the past, and in all such instances the organism attended to *relevant* ("already interpreted") situations (Tomasello, 2014). The outcome is that infants' iconic representations support goal-relevant attention, recognition, inference, anticipation, and learning about many different kinds of objects, people, and events, including some that she is not currently perceiving. But importantly, the infant is in all cases focused on the actual world, whether or not she is currently perceiving it. This contrasts with toddlers' abilities, as we shall see in the next chapter, to think, plan, and imagine potential actions and their likely outcomes, along with hypothetically possible external transformations.

3.2. Early Goal-Directed Actions and Decision-Making

In the first few months of life, infants use their iconically formatted cognitive representations to recognize and anticipate perceptual experiences and to learn many things about the world so represented, even though they are mostly not acting agentively on the world. But then, at about four or five months of age, their bodies begin to catch up, that is to say, they begin to use what they are perceiving and learning to act on the world in a goal-directed manner. This means that now their attention is drawn not only to situations that nature has determined are always relevant (bottom-up), but also to situations that they find relevant for their own agentive actions (top-down). So they now can learn not only contingencies between events in the world but also contingencies between their own actions and events in the world.

3.2.1. First Goal-Directed Actions

At around four to five months of age, infants begin to reach for external objects, typically grasping them and bringing them to the mouth. They quickly learn from such experiences how their actions produce effects in the world, even in highly novel circumstances. Thus, when Piaget (1952) tied one end of a string to a five-month-old's hand or foot and the other end to a mobile, she quickly discovered that her movements produced interesting effects and so repeated them many times (see also Rovee-Collier, 1999; Sloan et al., 2023). Human infants from four to five months of age thus have a propensity not only to attend, bottom-up, to deviations from their perceptual expectations but also to attend, top-down, to those aspects of their perceptual experience that are relevant to their goal-directed actions. If attention is the gateway to cognitive representation, infants from four to five months of age now have a new source of information for constructing their representations: top-down judgments about the relevance of events and situations to the agentive action they are performing or gearing up to perform. Learning based on top-down attention is again characteristic of a wide range of animal species, very likely from the earliest vertebrates.

3.2.2. Go/No-Go Decision-Making

The capacity for goal-directed action requires infants to make decisions about whether to execute an action in a particular situation. It is possible that they are making an either/or decision about which action to perform, but it is also possible that they are simply deciding whether to perform a particular action in that situation. This is the form of go/no-go decision-making that was very likely characteristic of the first goal-directed vertebrates. Thus, research with modern-day reptiles suggests that they "learn what stimulus to respond to rather than how to respond to a particular stimulus" (Wilkinson & Huber, 2012, p. 141).

At first blush, it would seem that infants do make either/or choices between alternatives. Thus, Hamlin et al. (2007) presented six-month-old infants with two stuffed animals, one of which had behaved more nicely than the other. Infants tended to touch or grab the nice animal, which could be taken as evidence of an either/or decision between the two options. But it is also possible that in their initial observation of the animals' behavior infants developed an attraction to the nicer animal, and as soon as they saw it, they went for it without comparing the relative values of the two different options. Under this interpretation, they are making a go/no-go decision for an attractor, not an either/or choice among alternatives.

Evidence for this interpretation comes from studies in which infants and toddlers have a prepotent tendency to go for a "wrong" option. The point is that if they succeed in overcoming this prepotent tendency, it suggests that they have attended to both alternatives and made an either/or decision. One example is action-based object permanence tasks. If a desired object is hidden under a single cloth, eight-month-old infants quickly remove the cloth and retrieve the object. But at this same age they often make the famous A-not-B error. This error occurs in a version of the task in which the infant is confronted with an object hidden under one of two cloths. After she finds it under cloth A, it is placed in plain sight under cloth B. In this two-cloth situation, infants often search for the hidden object under the cloth where they last found it (A), rather than where they last saw it disappear (B). They make this error through the end of early infancy, first searching reliably for the object in its new location (inhibiting any prepotent attraction to the first location) only as toddlers at around 11 months of age (Diamond, 1985; Marcovitch & Zelazo, 2006). The point is that the single-cloth task only requires the infant to make a go/no-go decision (to remove the cloth or not), whereas in the A-not-B task, she is confronted with an either/or decision between the two cloths, each of which is a salient alternative for good reason. Infants' behavior in detour tasks is similar. If a desired object is placed behind a transparent glass barrier, infants up to 11 months of age tend to just reach directly for the toy and bang into the glass (Diamond & Gilbert, 1989; Diamond, 1990), which implies that they are not choosing between the two alternative actions but simply seeing an opportunity to grasp an object and going for it. Toddlers, after 11 months of age, succeed in choosing the less salient alternative action.

The hypothesis is thus that young infants' actions are generated by a process of decision-making that simply determines whether to perform a particular action in the situation at hand: is this an opportunity for a particular goal-directed action? One might propose that the problem for infants is not decision-making but inhibitory control, and this would not be totally off-base. But either/or decision-making and inhibitory control go hand-in-hand in the sense that choosing among options means inhibiting the unchosen option. I would thus characterize the issue more broadly as infants not yet having an executive tier of functioning that can simulate alternative action possibilities and their likely outcomes, and so they do not yet have the possibility of either/or decision-making with proactive inhibition of unchosen behavioral alternatives. It is interesting that attempts to measure inhibition in infants before nine months of age almost all involve so-called delayed response tasks (e.g., Diamond, 1990), which only measure something like global inhibition of a single action and not selective (proactive) inhibition of one alternative versus another.

Evolutionarily, go/no-go decision-making is characteristic of contemporary reptiles (and, by hypothesis, early vertebrates). They, like young infants, fail both A-not-B tasks and detour tasks (see Tomasello, 2022a, for a review). This contrasts with the decision-making of many mammalian species who, like human toddlers, are skillful at either/or decision-making, including in both A-not-B tasks and detour tasks (MacLean et al., 2014). Suggestively, it has been observed in a number of experiments that mammals pause to look back and forth between alternative choices whereas, to my knowledge, human infants before nine months of age have not been reported to pause and look back and forth between alternatives in this same way.

3.2.3. Agency and Causality

If children are at some point going to formulate theories about how the world works, they are going to have to understand something about causality. Prototypically an understanding of causality means an understanding of forces external to oneself. But sometimes the term is also used to refer to experiencing the efficacy of one's own agentive actions. Because these two phenomena may be somewhat different, I refer to experiencing the efficacy of the self's agentive actions as s-causal to distinguish it from the understanding of external forces.

Experiencing the efficacy of self's agentive actions in s-causal terms is not confined to humans. Dickinson (2001) makes the case for laboratory rats. Briefly, if a rat acts and a result comes only after a several-second delay, they do not learn to perform that act for that result—presumably because it did not seem to s-cause it. Further, if a rat is in an environment in which its agentive act is followed immediately by a result, but in addition that result comes many times in the absence of any action, they again do not learn to perform that act for that result—presumably because that result seemed to occur independent of that act. Although, to my knowledge, experiments of this type have not been done with non-mammal vertebrates, my hypothesis is that the experience of an s-causal connection between one's acts and their results is an inherent part of being a goal-directed agent.

Millar and Watson (1979) found something similar in six- to eight-month-old human infants. When infants acted and then experienced an immediate result in the environment, they learned the contingency. But when the environmental result came only after a several-second delay, they did not learn the contingency. Relatedly, when infants act toward another person, they expect her to react more-or-less immediately and are confused if there is a several-second delay (Rochat et al., 1999). Tellingly, something similar is characteristic of infants' experience of causality when observing external events independent of their own action. In many different studies it has been found that infants at the same age as Millar and Watson's infants (~6 months) expect that when one animated object moves and contacts another, the recipient object will move as a result (Michotte's famous launching event). But they do not expect this result if there is a several-second delay between the initiating event and the result or if there is a lack of contact (see Saxe & Carey, 2006, for a review).

Several lines of evidence suggest a relation between infants' experience of s-causality in their own goal-directed actions and causality in external events. First is the fact that in both cases the connection is disrupted by a several-second delay between the instigating event and the consequent event. Second is the fact that both types of causal detection emerge at around the same age, namely, in the middle of the first year, suggesting a possible developmental relation. Indeed, one hypothesis is that, for infants, these two types of causal events are actually just two manifestations of the same phenomenon. That is, when infants themselves act and affect an object, this is itself a kind of launching event: the infant observes her hand make contact and move the object. White (2006) makes two other relevant observations that support this interpretation. First, when adults see something like a launching event, they experience the moving/initiating object as causing the recipient object to move, but they do not experience the recipient object as causing the initiating object to stop its motion (although this is just as available in the perceptual experience). Possibly, this derives from the fact that when we manipulate objects as agents, we typically do not focus on their effect on us. Second, White observes that we experience causality most directly and strongly in events involving inanimate objects when they concern the ways that we ourselves could affect objects through our actions, such things as "launching, entraining, pulling, and enforced disintegration ... because we can all kick, push, pull, and smash things from an early age" (p. 183).

But there remains the question of how infants understand the notion of causal *force*. It is possible that they experience their own actions as imparting some kind of s-causal force on the external world (they "feel" the energetic force of their own actions), as seems intuitive, though the evidence for such a conclusion is thin. The case for understanding external causal forces is thinner still, since all we have are infants' looking times. It is thus unclear how young infants understand either the s-causal efficacy of their own actions or external causal forces beyond spatial-temporal contingency. This lack of clarity contrasts markedly with the way that one- and two-year-old toddlers understand both self and external causal forces as well as logically structured paradigms of causal relations, as we shall see in the next chapter.

3.2.4. Infant Learning

Young infants' cognitive "priors"—their core cognitive representations in iconic format—structure their early learning about myriad specific objects, events, and relations in the world. They learn about things that draw their attention involuntarily, bottom-up, for example, when they learn that footsteps reliably signal adult arrival or when they learn novel facts about an object that participates in expectation-violating events (see Stahl & Feigenson, 2015, Studies 1–3 with 11-month-olds, whose results we might also expect in younger infants). Additionally, several-month-old infants also learn about the effects of their own actions on the world, top-down, as when they grasp and bang a spoon (versus a rope) on the table, or when they learn about the effects of their leg movements on experimentally constructed mobiles hanging above them (Sloan et al., 2023). And by six months of age, when infants see another person produce an interesting effect on an object, they often attempt to produce that effect on that object themselves (emulation learning; see Rochat, 2004, for a review of studies of infant learning in general). Let us call all of this attention-directed contingency learning.

Attention-directed contingency learning is also characteristic of many other vertebrate species, although the content of what is learned is of course very different across species. But new types of learning are on the horizon in human development. Specifically, as they develop a deeper understanding of both the s-causality of their own actions and the causal forces that structures events in the external world, one- and two-year-old toddlers will soon go beyond these straightforward types of attention-directed contingency learning to a more flexible kind of means/end exploratory learning—and indeed active hypothesis testing (see Stahl & Feigenson, 2015, Study 4, with 11-month-olds)—based on causal analyses of precisely what is causing what to happen in various kinds of actions and events.

3.3. Interacting in the Social World

People are special sorts of objects. They both do special sorts of things, and special sorts of things can be done to or with them. To make a cup move one pushes it, but to get an adult to do something one uses less direct means. And cups only move as a reaction to some force, whereas other people are constantly acting spontaneously in surprising ways, and these actions sometimes correspond to actions the child herself is performing or could perform. Moreover, people can be social partners with whom one may engage reciprocally. The point is that if children at some age are going to formulate theories about how other people work, they are going to have to understand something about agency, as a special form of animate causality. Since early vertebrates were not social creatures, most of infants' skills and knowledge for dealing with the social world almost certainly have evolutionary roots in more recent mammalian ancestors, who are generally social, including especially great apes, who have some special social and social-cognitive skills.

3.3.1. Understanding Others as Goal-Directed

Perhaps infants' most impressive social-cognitive skill is their ability to understand that the actions of others are, like theirs, often directed at particular objects. This understanding may be seen most clearly in the experimental paradigm pioneered by Woodward (1998). The general finding is that five- to nine-month-old infants who observe a human hand reaching toward an object expect that, on future occasions, that hand will continue reaching for that same object even if it has moved to a new location (rather than reaching to the old location). So they expect a person (presumed to be behind the hand) to continue reaching toward the same goal-object, which they do not expect a mechanical claw to do. This basic finding has been replicated with many task variations, including with great apes (Kano & Call, 2014).

It is striking that infants' understanding of the goal-directed actions of others emerges at the same basic age (~5 months) as their own first productions of goal-directed actions (e.g., in reaching for objects). Woodward, Sommerville, and colleagues have thus proposed that infants' experience of their own goal-directed actions is crucial for their understanding of others' goal-directed actions. As evidence for this proposal, Sommerville et al. (2005) provided three-month-old infants with extra experience in goal-directed action on objects by providing them with so-called sticky mittens that enabled them to better "grasp" and manipulate objects. They then tested them in the Woodward (1998) paradigm for understanding the goal-directed actions of others. As compared with a control group, infants who got this extra experience of reaching, "grasping," and manipulating objects were more adept at recognizing goal-directed reaching in others. Importantly, Gerson and Woodward (2014) tested the possibility that this effect was due merely to infants being able to observe goal-directed reaching by any agent: they compared "sticky mittens" infants to infants who observed similar amounts of sticky-mittens reaching and grasping by another person. They found that it was only infants' own production of goal-directed action that was effective in facilitating a precocious understanding of others' goal-directed actions. This is presumably because infants experience their own goals in a more direct way than they experience the goals of others. Indeed, from as early as two months of age infants can detect violations in which their own body does not move in accordance with their action goal (they are given experimentally manipulated video feedback about their leg motions), which of course is not possible when observing the actions of others externally (Rochat & Striano, 2000).

Woodward et al. (2009) review evidence from studies of older infants and toddlers to argue that this is a more general phenomenon: developing children use the experience of their own goal-directed actions to understand other types of actions (e.g., the pointing gesture). Spelke (2022; see also Woo et al., 2023) is skeptical of this analysis because it is not empirically the case that infants only recognize the specific actions of others, for example, jumping, after they themselves are able to perform that action. But a hypothesis can be formulated that is not so specific but still embodies the same general idea. The claim would be that young children understand the actions of others as organized or structured in the same way as theirs—for example, as goal-directed (or, later in ontogeny, intentional or metacognitive)—which enables them to learn to recognize novel actions that are organized or structured in this same way. In particular cases, such as reaching and grasping with sticky mittens, experiences of self-action may simply make others' similar actions and their underlying psychological processes more immediately salient, and this added salience is the basis of the experimental effect.

A variety of research paradigms—both behavioral and neuropsychological—suggest that, in fact, humans have a system of "common coding" for their own actions and those of others (see Prinz, 2012, for a review). Meltzoff (2005, 2007) argues that this is in place basically from birth, as evidenced by the fact that even neonates mimic the facial movements of others, which requires them to translate actions they can only observe visually into their own self-produced actions. He argues further that this imitative behavior indicates that infants experience other persons as "like me," and that this experience underpins their later social-cognitive capacities for understanding the mental states of others. Tomasello (2023) provides evolutionary support for the hypothesis of common coding, suggesting that adaptations for social learning and imitation require the individual to align not just her actions but also her psychological and intentional states with those of other agents. Looking across the spectrum of mammalian species, those that seem to be especially skillful at understanding the psychological states of others are those that are also skillful at imitating the actions of others (and at recognizing when they themselves are being imitated). Thus, most mammals are neither skillful imitators nor skillful mind readers, but great apes are both the best imitators and best mind readers among mammals, and this correlation is especially clear in humans, of course. Importantly, newborn chimpanzees engage in neonatal imitation of the same basic type as human infants (Myowa-Yamakoshi et al., 2004), and the only other mammalian infants to show this behavior are those of rhesus monkeys, who have some robust mindreading skills as well (see Rosati & Santos, 2016, for a review).

My view is thus that human infants, as great apes and imitators, are evolved to experience their own goal-directed actions and the goal-directed actions of others in the same basic terms, and, indeed, we shall see in the coming chapters that this applies, *mutatis mutandis*, to older children's intentional and metacognitive actions as well. This does not mean that children only understand specific actions and mental states of others after they have first experienced those exact same actions and mental states in themselves. But since first-person experience of one's own actions is privileged in containing information about intentional states in a way that third-person observation of others' actions does not, additional first-personal experience with particular actions (given processes of common coding) may facilitate attention to the similar intentional states of others as they perform similar actions.

3.3.2. Getting People to Do Things

For the first few months of life, infants do not do things to spur others into action beyond more-or-less reflexive or stimulus-driven crying when they are uncomfortable. But then they learn that they can agentively influence the actions of other people. In the same general manner that they learn to control the movements of a mobile to which their foot is connected by kicking, four-month-olds learn, for example, that touching their father's face makes him produce a raspberry sound or grabbing their mother's outstretched finger makes her produce tickling. By six to eight months, infants learn to ritualize some of these social behaviors and use them in a clearly goal-directed manner. Most commonly, infants who want to be picked up raise their arms while looking to the adult—suggesting that she insert her hands under their arms—and the adult often complies (Lock, 1978). Or infants who are being held but want to be put down often arch their backs and go partially rigid (ritualized from trying to wiggle free), and the adult often complies. Or infants who want their mother to set their baby swing back in motion kick their legs and shake their body to get the adult's attention, and again the adult often complies.

Piaget (1952) reports many similar examples. Infants are learning action-outcome pairings—in this case with animate beings—that they can actively use in a goal-directed manner in response to particular situations when they appear. Great ape infants and juveniles do many similar things both with humans who might be raising them and with conspecifics, for example, to initiate a certain type of play (Gómez, 2006; Tomasello & Call, 2019). Actions such as these that are aimed at social goals require the infant and nonhuman ape to understand at the very least that other animate beings spontaneously produce actions and that they can prompt (s-cause?) others to act in predictable ways through their own actions.

3.3.3. Engaging with Social Partners

All of this basic social cognition is shared among all great apes, including humans. But from two months of age, human infants do something unique: as they interact with adults, they express positive emotions in the species-unique behaviors of smiling and laughing, even from a distance, whereas other great apes do not smile or laugh in these ways. Trevarthen (1979) noted that there is also a kind of turn-taking in infants' and adults' interactions: first one acts while the other is passive, and then the reverse. Stern (1985) observed in these same kinds of interactions a kind of "emotional attunement" in which infants and mothers mirror one another's emotional intensity and valence, though often using different means; thus, the mother might express a positive emotion in her vocalization, and the infant might respond by smiling. Because these kinds of interactions are broadly communicative and reciprocal, Trevarthen has called them protoconversations.

The importance of this kind of emotional engagement for infants is clear if one abruptly terminates a protoconversation. Studies employing what has been called the "still face" paradigm have shown that if an adult suddenly presents an emotionless face in the middle of a protoconversation, infants begin showing various signs of upset (e.g., looking away, gnawing at their hands; Tronick, 1989). In addition, if the timing of the interaction is perturbed (in video set-ups in which the timing of mothers' responses is experimentally controlled as in Murray & Trevarthen, 1985, and Rochat et al., 1998), infants become upset as well. Great ape infants and their mothers do not engage in such protoconversations. Chimpanzee infants who have been raised by humans look at human faces in interesting ways and make various kinds of interesting facial expressions (Bard, 2012). But since apes do not smile or laugh like human infants, their emotional engagement with a human differs significantly from that which occurs in human mother-infant protoconversations. Hrdy (2006, 2016) has stressed that, given the unique way that humans raise their children cooperatively among many adults (i.e., unique among old-world primates, including great apes), human babies are faced with a much more demanding set of social challenges to obtain needed social support. Plausibly, this situation set the context for human infants to evolve especially powerful skills to connect psychologically with their mothers and other adults from a distance via species-unique emotional expressions such as smiling and laughing, as well as more extended protoconversations.

Almost all aspects of infants' cognition are shared with other animal species: their basic agentive organization and decision-making are shared with other vertebrates; their basic sociality is shared with other mammals; and their impressive skills of imitation and social cognition (perhaps due to common coding) are shared with other great apes. But the emotion-sharing characteristic of protoconversations would seem to be unique to the human species. Indeed, Tomasello (2019, 2020a) has proposed that human infants' unique propensity for sharing and aligning emotions with caregivers constitutes the underlying emotional substrate for later, more complex cognitive processes of shared agency and intentionality that, by all appearances, are also unique to the species (as we shall see in coming chapters).

3.4. Infants as Goal-Directed Agents

The earliest vertebrates on planet Earth foraged for highly mobile insects, requiring them to adapt to an ecology filled with unpredictabilities. They adapted by evolving a psychological architecture of goal-directed agency in which they attended to relevant situations and learned from them, resulting in perception-based iconic representations that guided their expectations and actions in novel situations.

The hypothesis I have explored here is that human infants begin life operating in this same general manner as goal-directed agents. In the spirit of evolutionary developmental biology (Evo-Devo) one may ask why it is adaptive for human infant psychology to be organized in this way. The answer very likely is indirect: this organization simply represents a necessary first step in building up a more complex psychological architecture, and it persists for so long (nine months) because it can. That is, like other mammals and primates, humans have evolved to develop slowly—partly to maximize time for learning—and this works because parents and other adults act as agents for infants to provide them with what they need.

The evidence reviewed here suggests that indeed human infants begin life as (still very immature) goal-directed agents. Nature has therefore equipped them with perception-based iconic representations that structure their bottom-up attention, expectations, and learning. After a few months, infants begin to actively pursue behavioral goals, and this activity now structures their top-down attention, expectations, and learning. Infants' decision-making in these activities is of the simple go/no-go variety: they recognize a situation and decide whether to act in some manner, without any executive oversight or control (except for a reactive kind of global inhibition when necessary). As they are attending to the physical world and acting in it, infants are learning contingencies that hold both between their actions and changes in the world and between changes in the world independent of their actions (what I am calling "attention-directed learning"). It is unclear whether they conceptualize any of these contingencies specifically in terms of causal forces.

This is generally similar to the way that other vertebrates act in the physical world. But some of infants' more specific cognitive capacities have a more recent evolutionary origin, for example, their great ape fascination with grasping and manipulating objects. Also of more recent evolutionary origin are their capacities for social cognition and interaction. Of special importance for infant social-cognitive development are (i) the system of common coding via which great apes perceive and understand the actions of the self and others in similar terms (very likely evolved in the context of great ape social learning); and (ii) the processes of emotion sharing and attunement in protoconversations (very likely evolved in the context of infant–adult interaction in early human cooperative childcare). These special adaptations set the stage both for toddlers' later understanding of others as intentional and metacognitive agents and for their acting in joint agencies with them.

Overall, much of what we know about infant cognition has been discovered using methods that exploit infants' tendency to attend especially intently to unexpected events. When these methods were first introduced, the claim was that they enabled us, as scientists, to overcome the inconvenient fact that infants are incapable of expressing their knowledge in action or language. The assumption—either implicit or explicit—was that the knowledge that infants' looking behavior revealed was knowledge of the same general type possessed by older children and adults; we had simply found a way to observe and measure it. But another possibility is that infants' knowledge of the world is of a specific type that differs in systematic ways from the knowledge of older children and adults.

The proposal here is that human infants' knowledge is of a special nature because it is aimed exclusively at the actual world as they experience it. That is, they begin with some built-in iconic representations of the world, and they learn others. They make simple inferences about what is the case in the actual world—even if they are not currently perceiving it—and they anticipate what will happen next, with learning focused on potentially observable sequences of events (not on unobservable causal or intentional forces underlying events and actions). Infants are not living wholly in the here and now—they can recall and anticipate—but neither are they living in a world of possibilities in which they proactively think, plan, and intentionally organize their actions based on the imaginative evocation of cognitive content on an executive tier (e.g., in making either/or choices among imagined actions and their likely outcomes), which will not be possible until after nine months of age. As compared with older children, infants are cognitively distinct because they are living and learning solely in the *actual* world as they experience it.

II
TODDLERHOOD

The adaptation of intelligence . . . depends as much on progressive internal coordinations as on information acquired through experience.
                                                                    Jean Piaget

NB: The iconic representations characteristic of young infants—as established by looking-time studies—persist and are used throughout later developmental periods into adulthood. But because I want to focus here on toddlers' newest, cutting-edge cognitive capacities, I concentrate in this chapter on studies in which toddlers use their representations imaginatively to think, plan, and make active behavioral decisions.

4
Intentional Agency and Imaginative Representations

At around 9 to 12 months of age, infants transform physically into toddlers, in the literal sense that they are beginning to locomote more independently, perhaps even starting to walk. Since independent locomotion creates new challenges for behavioral decision-making, it is perhaps no accident that at this same age infants' goal-directed agency transforms into toddlers' intentional agency, which ushers in a whole new mode of cognitive functioning.

Most basically, toddlers begin to make either/or decisions between imagined behavioral options with their imagined environmental outcomes. This requires cognitive representations employed not just to recognize objects and events in the actual world, as are infants', but to imagine potential actions and outcomes in the possible worlds of thinking and planning. In this way of operating, the output of a decision is not an action but an intention to act, which enables the compilation of hierarchically embedded intentions before acting (e.g., planning to remove an obstacle on the way to a goal). Such proactive thinking and planning require that toddlers operate with some new executive processes, including proactive inhibitory control (to suppress unchosen behavioral options before acting), hypothesis-directed exploratory learning, and constructive thinking and re-representation.

The working hypothesis of this chapter is that what enables this new manner of agentive functioning is the emergence of a new organizational architecture, namely, one that includes a single tier of executive supervision and control. The resulting new form of intentional agency and its underlying architecture have their evolutionary roots in the earliest mammals and are characteristic today of species such as squirrels, rats, and domestic dogs (see Tomasello, 2022a, for a review).

But in addition, as great apes, toddlers are also beginning to attribute agentive powers to outside objects and people, based on their newfound executive experience of their own intentional actions (given a system of "common coding"). These attributions create for toddlers two new dimensions of

Agency and Cognitive Development. Michael Tomasello, Oxford University Press. © Michael Tomasello 2024.
DOI: 10.1093/9780191998294.003.0004

experience based on their new understanding of the world in terms of (i) the causal forces underlying physical events and (ii) the intentional forces underlying agentive actions. Understanding these forces—and how one might intervene to affect them—equips and empowers young toddlers to construct their first hypotheses and theories about *why* things happen as they do. Construction of these hypotheses and theories relies not only on toddlers' new understanding of causality and intentionality but also on their agentive skills of constructive thinking and executive re-representation.

4.1. Intentional Action and Decision-Making

Six-month-old infants recognize a situation and decide whether to perform a given action. Nine-month-old and older toddlers assess a situation—sometimes including elements of its causal and/or intentional structure—and decide which action(s) to perform, a decision that quite often requires thinking, planning, and proactive inhibitory control. As a result of this more complex way of relating to and acting in the world, toddlers' learning goes beyond simple contingencies to include means/ends (causal) analysis of how actions s-cause their results as well as how external events cause one another. Toddlers could not act in any of these ways equipped only with infants' perception-based iconic representations of the actual world; they need imaginative representations of some possible worlds.

4.1.1. Thinking, Planning, and Imaginative Representations

When Piaget (1952) interrupted his seven-month-old daughter's reach for a toy by placing a pillow in the way, she seemed perplexed and quit reaching. But at nine months of age, she reacted by deliberately grasping the pillow and setting it aside, immediately then procuring the toy. Removal of the pillow in such situations represents a hierarchically embedded sub-procedure in the main act of obtaining the toy. It is an intentional action because the removal of the pillow is done *in order to* obtain the toy. Also in this same category, toddlers soon begin using tools of various kinds for various ends, for example, using a stick to rake in out-of-reach objects. Acts of tool use are again hierarchically embedded intentional acts because the manipulation of the tool is done only in order to attain some other end. To organize intentional actions of these kinds, toddlers must proactively plan their actions in an executive-tier workspace (what some may call working memory).

Some months later toddlers begin to solve more complex problems by constructive thinking. For example, a toddler might want to open a door but an object is directly in front of it (Piaget, 1952). A 12-month-old would likely pull the door, watch it collide with the object, and only then try something new. But the 18-month-old imagines the door swinging open and being impeded by the object, and so removes the object before actually pulling the door. This would seem to involve constructive thinking (in the form of "insight") about the causal structure of the situation and planning one's actions accordingly. In the studies of Bauer et al. (1999), young 2-year-olds thought and planned especially deeply and efficiently, for example, planning a two-step strategy to construct a toy that can produce an effect by imagining ahead of time various action-outcome sequences (i.e., they had to place a small block inside one half of a toy barrel before snapping on the other half of the barrel—so that they could then shake it to make a rattling sound). Progress in planning and problem-solving continues unabated from ages two to three years.

To engage in proactive thinking and planning of this kind, toddlers need to use their cognitive representations in new ways. Whereas infants operate with iconic representations activated by perceptual experiences, toddlers must be able to voluntarily evoke and manipulate representations imaginatively. One hypothesis is that the core iconic representations from infancy persist during toddlerhood (indeed throughout the lifespan), but representations that are learned for new cognitive content—whether learned during infancy or later—operate in this new imaginative format from nine months of age forward. These new representations are still based in perception—what is imagined is possible perceptions, or schematic versions thereof, based on past perceptions—it is just that they now may be creatively evoked and manipulated in the absence of immediate perceptual stimulation. This is what the new architecture of intentional agency both requires and empowers. It also supports toddlers' emerging capacities for pretense, which are such an important part of their life from 18 months of age onwards (Harris, 2000).

4.1.2. Either/Or Decision-Making and Inhibitory Control

In contrast to infants, toddlers make either/or behavioral decisions in which they imagine behavioral options with their likely outcomes and then choose one before acting. This is what Berkman et al. (2017) call "value-based choice," in which the preferred option is increased in value, and/or the less preferred option is decreased in value, relative to the other(s). Leahy and Carey (2020) are skeptical that toddlers engage in this kind of either/or decision-making in which they simultaneously consider two possible alternatives. In their analysis, the problem is that children do not have a representation of the modal concept "possibility" until something like four years of age. They believe that toddlers operate only with "minimal representations," in which they consider alternatives sequentially, making a go-no-go decision about each separately and then seeing if it works in actual behavior before trying another—what they call "sequential guessing."

But the theoretical framework within which Leahy and Carey analyze toddlers' decision-making is a kind of "language of thought" in which each alternative is represented as a proposition about the external world that is either true or false. For example, facing an adult preparing to drop a toy down an upside-down Y-shaped tube, the toddler, in this analysis, cannot hold in mind at the same time two alternative causal propositions about the world, each of which may be either true or false (e.g., the toy will come out either to the right or to the left). This may very well be the case. But what I am proposing is not the toddler simultaneously keeping in mind two incompatible causal possibilities—with an indeterminant causal outcome at the moment of choice—but rather the toddler simulating two potential behavioral options and their likely outcomes, and then choosing the option with the best likely outcome based on some form of Bayesian probabilistic inference (see Alderete & Xu, 2023, and Section 4.2.3). This was seemingly the process when the chimpanzees studied by Engelmann et al. (2021) were uncertain about which of two boxes contained a reward, and so pulled them both in simultaneously (and then searched only until they found the one reward). And something similar was at work when the chimpanzees studied by Engelmann et al. (2023) protected both available pieces of food from a competitor when they did not know which he would be trying to steal (but only one piece of food when they knew his target).

I believe one can see the origins of toddlers' either/or decision-making already in their behavior in the two-cloth object permanence task, as noted in the previous chapter. Soon after nine months of age toddlers begin to solve this task without making the A-not-B error: they choose which of the two cloths is likely concealing the desired object and choose that one. This value-based choice involves a more flexible form of inhibitory control than the simple global inhibition characteristic of infants. As toddlers are comparing the behavioral options, choosing the cloth where the object just disappeared means proactively suppressing the prepotent tendency to choose the cloth where the object was previously found. In support of this interpretation, much research shows that toddlers' ability to make choices in this manner correlates strongly with their performance in tasks measuring inhibitory control (Marcovitch & Zelazo, 2006). Moreover, either/or comparisons of this kind should take time to execute, and Kim et al. (2020) found that when 12- and 24-month-old toddlers are faced with more uncertainty in their potential choices, they take more time to decide. In general, toddlers seem to be making proactive either/or decisions involving processes of executive inhibitory control.

Perhaps even clearer evidence for this kind of decision-making comes during this same age range as toddlers make decisions in so-called opt-out tasks requiring them to compare options before choosing. A number of mammalian species—including dolphins, rats, and many nonhuman primates—have been confronted with a choice between an easy-to-obtain low reward and a more difficult-to-obtain high reward. When chances of obtaining the high reward are high, individuals will go for that; but when chances of obtaining the high reward are low, individuals often opt out and go for the easy-to-obtain low reward. Goupil et al. (2016) tested 20-month-olds in a situation with this logic (the opt-out response in this case was to request adult help) and found that toddlers made efficient choices. Further, Call and Carpenter (2001) found that when 30-month-olds felt uncertain about a decision, they actively sought more information to try to make a better decision, again showing the ability to comparatively evaluate alternative possible actions.

But the strongest evidence comes from an experimental paradigm typically characterized as measuring inhibitory control or executive function. The situation is slightly different from uncertainty monitoring in that the costs and risks of both possible choices are clear (often with one having a kind of prepotent attraction). Thus, Herrmann et al. (2015) confronted 36-month-olds with a spatial discounting task in which the child first spied a nearby small reward and then a farther-off large reward, and they were shown that going for one meant forsaking the other. They had to compare the two situations and make a choice before acting, which prevented a sequential guessing strategy. In a similar task, toddlers had to choose one of two behavioral strategies given that the situation had noticeably changed, which meant inhibiting a previously successful action in favor of a new one demanded by a changed situation (again they had to choose before acting so a sequential guessing strategy was not possible). In both of these tasks, toddlers were generally successful (equally as good as chimpanzees but not as good as six-year-olds). Toddlers' behavior in all these tasks thus suggest either/or, value-based choices between two simultaneously available courses of action as they imagine them.

Such value-based decision-making among simultaneously available options cannot take place in creatures that operate as a simple goal-directed control system operating only with goals, actions, and attention. Rather, it requires an additional executive tier of monitoring and control to regulate the process. Intentional agency of this type is characteristic of most mammals, with much research showing that species such as squirrels and rats cognitively simulate problems, evaluate potential actions, and then make either/or decisions for one action plan (inhibiting unchosen options) (see Tomasello, 2022a, for a review).

4.1.3. The Executive Tier

How might we best conceptualize this executively organized and regulated manner of agentive functioning? In cognitive and developmental psychology, the term "executive function" refers to a dizzying array of diverse processes, often defined within particular research paradigms, leading to a proliferation of theoretical constructs focused on specific tasks. In developmental psychology, executive function can refer to such diverse phenomena as behavioral inhibition, cognitive inhibition, inhibitory control, self-control, effortful control, proactive executive function, continuous monitoring, working memory, self-regulation, emotion regulation, attention regulation, attentional control, attention shifting, cognitive flexibility, set-shifting, task switching, and others. Many researchers have bemoaned the plethora of terminological jargon in the field, and some have doubted the psychological reality of this menagerie of constructs (e.g., Doebel, 2020).

Diamond (2013) attempted to bring order to this menagerie by proposing a tripartite typology of basic processes: (i) Inhibition (e.g., inhibitory control, self-control, behavioral inhibition, emotion regulation, etc.); (ii) Working Memory (i.e., the ability to hold information in mind and mentally work with it in various ways); and (iii) Cognitive Flexibility (e.g., attention shifting, set-shifting, mental flexibility, etc.). Although this typology has proven useful in identifying individual differences in executive function as they relate to developmental outcomes such as school achievement and emotional adjustment, from a theoretical standpoint, the types in this typology are very diverse: "inhibition" is a basic psychological process, "working memory" is a cognitive workspace within which psychological processes operate, and "cognitive flexibility" is a trait that people or processes possess.

In the current model executive processes are not just an unorganized collection of independent mechanisms; they are all a part of, or a specific application of, a regulatory system evolved to monitor and control behavioral decision-making and action. That is to say, executive processes all function within an executive control system whose goal is to facilitate the functioning of a behavioral control system. The agent pursues this executive goal on an executive tier of operation by simulating, monitoring, and controlling its attention and action on the behavioral tier of functioning. Specific executive processes, as they are currently defined in the field, operate as special applications of this general system (or components of it) in particular behavioral contexts (see Chapter 6 for some specific examples). This hierarchical model bears some resemblance to that of Zelazo (2004, 2015)—in ways also explicated in Chapter 6—but his model is focused mainly on explaining children's conscious access to various aspects of their psychological functioning, whereas the current model focuses specifically on the monitoring and control of attention and action in the process of behavioral decision-making.

4.2. Understanding Causality

In addition to these executively structured ways of making decisions and regulating actions, toddlers also construct some new dimensions of experience that enable them, for the first time, to identify not just what is happening but *why* it is happening. They do this by attributing to external entities underlying causal and intentional forces analogous to those structuring their own intentional actions. The attribution of causal and intentional forces to entities in the external world enables toddlers for the first time to construct theories—via processes of constructive thinking and re-representation—about why things are happening in the ways they are happening and to actively test relevant hypotheses. When the explanatory target is conceptualized in its most general terms—namely, the causal interactions of objects (naïve physics) and the intentional interactions of people (naïve psychology)—these theories are often called "framework theories" (Wellman & Gelman, 1992). Although they emanate from somewhat different evolutionary and developmental sources, toddlers' understanding of causality and intentionality are both grounded in children's executive experience of their own agentive actions. I deal with each of these in turn—causality in this section and intentionality in the next.

4.2.1. Making Things Happen via Intermediaries

David Hume famously argued that when we see one billiard ball smashing into and launching another billiard ball, we do not just see the temporal contiguity of two events, but rather we see one event "forcing" the other to happen by some kind of energetic transfer. Hume reasoned that since this force is not directly observed—we only see one event followed by another—observers must be attributing it to the initiating ball's action themselves. This unobserved force explains why the first event *causes* the second. When six-month-old infants are confronted with different launching events, they are clearly expecting a spatial-temporal structure consistent with the adult notion of causality. But it is unclear whether they have any expectations about an external causal *force*. Where does this notion of external causal force come from? And how does it support more complex inferences about causal relations?

As noted in the previous chapter, the causal force with which infants are most intimately familiar, from at least five months of age, is their own agentive acts. Beginning with Hume's Scottish contemporary, Thomas Reid, and continuing through a number of others, some theorists have argued that this is in fact the only causal force that humans can experience directly. In a review of young children's causal understanding, Saxe and Carey (2006, p. 147) quote the philosopher Maine de Biran: "A being who has never made an effort would not in fact have any idea of power, nor, as a result, any idea of efficient cause. He would see one movement succeed another . . . but he would be unable to conceive, or apply to this sequence of movements, the idea of efficient cause or acting force." White (2006, p. 173) stresses that when we ourselves cause something to happen through our physical actions we can actually perceive the pressure of our hand or body against the object: "Actions on objects haptically perceived (supplemented, of course, by visual and auditory information) anchor us in reality, and give us our most fundamental sense of what it means to generate an effect in something else."

But a causal event is prototypically one that happens in the external world independent of the self's actions. Infants' expectations about external launching events are a start in that direction, though, again, it is unclear if infants attribute to the launching event any sense of force. An interesting hypothesis is that only some species—and humans only after some point in ontogeny—understand external causal forces, and those are the species that are intentional agents adapted for making and using tools. Thus, as toddlers begin to use simple tools around their first birthdays, they themselves supply the force to the tool that causes effects in the environment. But it only works if the tool has certain properties that make it causally effective in the situation, for instance, it must be long enough and rigid enough to rake in an out-of-reach object. Using tools effectively thus results both from force applied by the toddler and by her understanding of the causal relation between the tool and the problem substrate. A good test of toddlers' understanding of the causal relation between tool and problem substrate is thus so-called tool choice. Wobber et al. (2013) presented two-year-old toddlers (and great apes) with a problem and a set of tools, some of which were appropriate for the problem and some of which were not. Both the toddlers and the great apes were able to effectively assess the way that different tools would or would not cause a result and so choose the most appropriate tool for the problem at hand—before actually using any of the tools physically.

In addition to tools, toddlers also understand the intentional actions of others as causing effects in the world. Thus, Hauf et al. (2004) had 12- and 18-month-old toddlers observe an adult perform a sequence of actions on a toy bear, only some of which produced a distinctive noise. When given the toy bear themselves, toddlers reproduced more quickly and more often those actions that produced the interesting effect (see Elsner, 2007, for a review of this and similar studies). If toddlers understand the intentional actions of others on analogy to their own (via common coding), one interpretation of this and similar studies is that the toddlers see the action of the other as causing the effect; otherwise, why would they think that their using the same action on this object would produce the same effect? Further evidence for this interpretation is the fact that at around 9 to 12 months of age, toddlers begin making intentional attempts to get others to cause effects in the world through various forms of intentional gestural communication. For example, they get an adult to transport an object to themselves by pointing to it and whining, or to fix a broken toy by showing it to an adult pleadingly (see next chapter for more detail).

Bates (1979) points out the deep parallels between toddlers' use of hierarchically embedded sub-procedures in their intentional actions and both their use of tools as causal intermediaries and their use of communicative gestures as causal intermediaries. In all three cases, the individual uses something—an action, tool, or gesture—as an intermediary that they hope will cause a desired effect either in some physical object or some animate agent. Among mammals, only great apes use both tools and gestures intentionally and flexibly as sub-procedures, and it is only after nine months that human children do these things as well. My hypothesis is that executive monitoring these various ways of using causal intermediaries—accompanied by constructive thinking and re-representation—leads to a greater understanding of the causal role of the intermediary itself, which paves the way for both apes and toddlers to attribute to external objects and events the "same" causal force that they experience in their own intentionally structured actions.

4.2.2. Understanding External Causes

In addition to the use of tools and gestures as causal intermediaries, nonhuman great apes can, in some contexts, understand causal forces that operate totally independently of their own actions. For example, in one study apes attempted to identify which one of several opaque bottles contained juice (they could choose only one). They quickly alighted upon the strategy of picking up the bottles one by one to test for their weight, and as soon as they found a heavy one chose it. In contrast, if the bottles all weighed the same but the bottle with juice was painted red, they found it extremely difficult to associate the red color with the presence of juice across trials. They thus understood, in some sense, that the extra weight of the bottle was caused by the juice (Hanus & Call, 2008). In another study in which they could not act on the objects at all, chimpanzees inferred that when one end of a balance beam tilted down it meant that the opaque cup on its end contained a banana (which they did not infer if a human pushed that end of the balance beam down), indicating an understanding that heavy things exert a downward causal force (Hanus & Call, 2011). Although the looking time studies with infants may be seen as investigating this same question of external causality, because the dependent measure is only looking it is not clear if they establish an understanding of causal force.

The most directly relevant studies with toddlers are those using the famous blicket detector machine. For example, Gopnik et al. (2001) presented two-, three-, and four-year-old children with a blicket machine and told them that blickets make the machine go. They then showed them various events in which different kinds of blocks were placed on the machine, and the question was which ones of them were blickets. For example, the child might see that A and B together make the machine go but B by itself does not, leading to the inference that A must be a causally effective blicket. Children of all three ages were reasonably competent in this task. Even two-year-olds could both answer questions appropriately and also physically intervene appropriately to start the machine by placing the appropriate block on it and stop the machine by removing the appropriate block. Focusing on even younger children, Sobel and Kirkham (2006) found that 19- to 24-month-old toddlers were competent in many of these same ways (see also Muentener & Schulz, 2014). Focusing on behavioral interventions, Bonawitz et al. (2010) found that after 24-month-olds observed a novel causal event in which one object contacted another and caused a reaction, they then imagined and so predicted what would happen if they themselves intervened in a particular way (the prediction evidenced both by their actions and by their looking in advance to the anticipated result; see also Meltzoff et al., 2012). Toddlers thus understand many novel events as causal before acting on them, which enables them to manipulate causes themselves to get desired effects. When great apes are tested in paradigms similar to this one, they make very similar causal inferences and predictions (e.g., Völter et al., 2016; Tennie et al., 2019).

Toddlers have thus begun the process of understanding *why* events in the physical world happen, that is, what causes or forces them to happen (see Goddu & Gopnik, 2024, for a recent comprehensive review). Beyond young infants' expectations of contingencies, toddlers' tool choice, reproduction of others' causal actions, and causal interventions suggest an understanding of the underlying causal *forces* in physical events, possibly on analogy to the internally sensed force of their own actions as they operate through an intermediary to force external effects. Still further considerations support this hypothesis.

4.2.3. The Logical Structure of Causal Inferences

Toddlers' causal inferences are not isolated, independent events; they are organized into logical, or quasi-logical, paradigms. For example, consider the situation in which a toddler (or an ape) chooses a tool that is causally appropriate for a task (as in Wobber et al., 2013). A causal understanding of the situation generates creative inferences organized into logical paradigms. Thus, in choosing a tool the toddler (or ape) infers such things as "if a tool with property A is used, then B will/must happen as a result." Then, actually using the tool completes the syllogism: (i) if A is used, then B happens; (ii) A is used; (iii) therefore B will/must happen. This form of thinking represents a kind of proto-modus ponens logical paradigm.

Or consider a more complex case. Call (2004) showed a chimpanzee a piece of food, which was then hidden in one of two cups (and the ape knew, from training, that it was in one or the other). In the simplest case, he shook the cup with the food, it made a noise, and the ape inferred that the food was inside. In a more complex condition, Call shook the empty cup, which produced only silence. In this case, to find the food the chimpanzee had to make a more complex set of interconnected inferences such as: (i) if food were inside the shaking cup, then it would make noise; (ii) the shaking cup is silent (not noisy); (iii) therefore, the shaking cup must be empty (not contain food). This represents a kind of proto-modus tollens inferential paradigm in which the absence of an particular effect implies the absence of a particular cause. The chimpanzees made this chain of inferences, but then also made an additional one: they combined this chain of inferences about the causality of noisemaking with their pre-existing knowledge that the food was in one of the two cups to locate the food in the *other*, non-shaken cup. This exclusion inference in this inferential context represents a kind of proto-disjunctive syllogism.

This basic experimental design was used by Mody and Carey (2016) with 23-month-old human toddlers, and they too were successful. However, based on further studies in which apes and children failed, Leahy and Carey (2020; see also Leahy, 2023) do not believe that either species is thinking in terms of a disjunctive syllogism. They believe that apes and toddlers can perform well in this task with, again, a sequential guessing strategy that does not involve "or" as a logical operator. To demonstrate an understanding of the logical "or," children would need to be using it propositionally, as a logical operator specifying that one alternative proposition was true and the other false. The newer studies require children to do this in more complex problem situations in which they must compare possible outcomes among multiple causal sequences simultaneously. Toddlers fail these, and so Leahy and Carey are skeptical that either they or the apes are using a logical "or," and indeed they argue that children will get to fully logical thinking only as they master the appropriate words in language, which is not until around four years of age. But I would argue, again, that while this may be a valid point of view for strictly logical concepts involving the truth conditions of causal propositions about the external world, apes and toddlers do not need to do that in either the original task or their everyday decision-making. They simply need to compare the likely outcomes of different behavioral decisions and choose among them based on a paradigm of probabilistic inferences—as did children in the recent study of Alderete and Xu (2023) in which three-year-olds were quite competent in tasks analogous to those of Leahy (2023) if they were structured probabilistically.

It may very well be that representing logical structure with linguistic symbols generates a new form of propositional understanding. Nevertheless, toddlers' and apes' thinking is already structured by a logic; it is just not a formal, propositional logic but rather a logic of causality. That is, logical inferences structured into paradigms do not come into existence with the advent of language, but they already characterize toddlers' inferences about causal events (language then inheriting this structure). The key insight is that causal thinking just is logical thinking. Bermudez (2003) argues that the thinking of apes and children in the experiments just described is logical because the logical paradigms instantiated—proto-modus ponens, proto-modus tollens, and proto-disjunctive syllogism—rest on the two pillars of classical logic: (i) the conditional (if... then ...) and (ii) a kind of negation. It is just that causal inferences are material (not formal) conditionals. Thus, the inference that if one thing happens then another will/must happen constitutes a kind of proto-conditional. A kind of proto-negation is based on what logicians call contraries: exclusionary opposites on a scale such as presence-absence, noise-silence, safety-danger, success-failure. If we assume that toddlers and apes understand polar opposites such as these as indeed mutually exclusive—for example, if something is absent it cannot be present, or if it makes noise, it cannot be silent—then this could be a much simpler basis for the negation operation. Thus, in the task of Call (2004) and Mody and Carey (2016) the apes and toddlers know that silence indicates an empty cup, noise indicates a full cup, and a single cup cannot be both—and this guides their action.

Taken together, the proto-conditional (if... then ...) and proto-negation (contraries) operations can structure all of the most basic paradigms of human logical reasoning, just in proto form as applied to the causal structure of events in the world. The claim is thus that great apes and toddlers are using this kind of proto-logic to simulate or make causal inferences about what has happened previously or what might happen next. But now the question arises where does this logical structuring of causal inferences come from? My hypothesis once again is that it comes from the agentive organization of the child's own actions understood as s-causally effective as they are executively monitored—yielding a kind of logic of action. Thus, the s-causal understanding of one's own actions as applied to external events by toddlers yields such inferences/implications (➜) as (given an understanding that A causes B):

- Act A ➜ Effect B (basis of Modus Ponens);
- Effect B absent ➜ Act A absent or ineffective (basis of Modus Tollens);
- Act A absent ➜ Effect B absent (basis of affirming the antecedent fallacy);
- (if only one of two acts, A or X, can cause effect B) Act A => absence of Effect B ➜ Act X => Effect B (basis of disjunctive syllogism).

These kinds of inferences are made on the executive tier about one's own goal-directed actions and their s-causal effects, as characteristic of mammals in general, based simply on the way that agentive action works. Adaptations for the use of causal intermediaries such as tools and gestures—when executively regulated with re-representation—then support toddlers' attributions of external causal forces.

4.2.4. Hypothesis-Directed Learning and Re-representation

With the emergence of an executive tier of agentive architecture, toddlers become capable of thinking, planning, and making value-based decisions in the production of intentionally organized actions imaginatively and proactively. In combination with their new understanding of both self and external causality, this new manner of functioning enables some new forms of learning.

Most basically, the planning of intentional actions—along with the subsequent executive monitoring of action production and its causal results—involves a kind of causal (means-ends) analysis or "blame assignment" of what factors are responsible for what results. Toddlers are thus able to go beyond the attention-directed contingency learning of infants to assess "what causes what" when they act on the world and so to learn how to produce or predict particular effects flexibly in novel situations. Thus, as they are using tools toddlers are beginning to understand what part is being played by their own action production and what part is being played by the causal relations between tool and problem substrate. This understanding of the causality both of their own actions and of external events means that toddlers can distinguish between cases in which (i) they intended an act to cause a result but they failed to perform the act adequately, in which case no causal inference about the world is implied, and (ii) they executed an act perfectly, but the environmental situation did not change in the expected way, in which case further examination of relevant external causes is in order. I know of no experimental research that directly supports this analysis for toddlers' physical actions, but in their communication 18- to 30-month-old toddlers adjust to a failed communicative attempt differently depending on what they diagnose as the cause of their failure (e.g., they articulated poorly, the listener misunderstood them, the listener does not want to comply, etc.; Grosse et al, 2010). This kind of learning is only possible for organisms that executively monitor their intentional actions with a causal mindset.

In addition, there is another innovation in learning processes made possible by the executive tier: hypothesis-directed learning. Whereas infants learn contingencies between actions and results or between external events, toddlers proactively explore such contingencies, again employing a kind of causal analysis or blame assignment to identify "what causes what," both with respect to their own actions and with respect to events in the environment. Thus, starting at around 12 months of age toddlers begin to perform acts just to see what outcome might result, which often leads to the discovery of new means toward desired ends and/or to the discovery of new causal relations in the environment (Piaget, 1952). For example, 12-month-olds do things like throw a toy from the highchair in order to see how it bounces, push over the planter to see how the dirt spills out, and pour water onto the cat to see what it does. This exploration can also be prompted by surprising events independent of the child's actions. Thus, in the study of Stahl and Feigenson (2015, Study 4), when 11-month-olds experienced an event that violated basic principles of object motion, they not only paid special attention, but they also took the object and dropped or banged it in various ways to actively explore hypotheses about the causal structure of this surprising event. These proactive forms of hypothesis-directed exploratory learning—which are also evident in the studies of causal learning cited above (e.g., Gopnik et al., 2001, and Sobel & Kirkham, 2006)—represent toddlers' earliest attempts to use their own intentional actions to test theories or hypotheses about why the world works as it does (see Povinelli & Dunphy-Lelii, 2001; Hanus & Call, 2008; Völter et al., 2016, for some related findings in great apes).

Toddlers' logical paradigms of causal inferences about the physical world also structure processes of constructive thinking and re-representation. That is to say, as they are learning about the physical world, toddlers are actively constructing theories that relate various specific causal relations to one another. For example, suppose a toddler learns that she may use a large spoon to scoop up dirt. On another occasion she learns that some toy animals are too large to fit through a hole in a Plexiglas panel, and that others will only fit through if they are turned in a specific manner. Now she is faced with a problem of some dirt she wants to scoop up on the other side of a Plexiglas panel with a small-ish hole, with several spoons of different sizes in the dirt. Using processes of constructive thinking, she might formulate the hypothesis that perhaps a smaller spoon is best for the task or that if she uses a larger spoon it must be rotated to get through the hole. She formulates these hypotheses even though she has previously only scooped up dirt with a large spoon, with no experience of a spoon being too large to fit through a hole in a panel or turning a spoon to make it fit through a small opening. These are thus creative hypotheses formulated by processes of constructive thinking and structured by the paradigms of quasi-logical inferences described above, for example, "if a large spoon is used, it will be blocked (not work)," and "if a large spoon is turned horizontally, it will work (not be blocked)." Assuming that one of these hypotheses produces a successful result, the outcome will be a new theoretical understanding of the spatial relations among objects and apertures, which will then be re-represented appropriately.

All these processes of constructive thinking, re-representation, and theory formation require an executive tier of functioning to provide the cognitive workspace (executive working memory) necessary for bringing together and coordinating disparate experiences. New representations may then potentially be analogized to other situations of the same type, again using the executive-tier workspace.

4.3. Understanding Intentionality

In addition to forming theories about why things happen as they do in the physical world, toddlers also form theories about why agents act as they do in the social world. To do this they need to understand others as intentional agents like themselves. We saw in the last chapter that infants begin to understand others as goal-directed agents at around the same time that they begin to perform goal-directed actions themselves (i.e., at around 5 months). Based on the theory of common coding, we might thus expect that as toddlers begin acting as intentional agents—guided by thinking, planning, and either/or decision-making—they begin to understand others as intentional agents who work in this same way.

4.3.1. Understanding Others' Goals/Intentions and Attention/Knowledge

Much experimental evidence suggests that great apes are not only intentional agents themselves, but they also understand others as intentional agents (Call & Tomasello, 2008). Our question here is whether, as they are becoming intentional agents themselves at around 9–12 months of age, human toddlers also begin to understand others as intentional agents as well, that is, as agents who operate with: (i) internally represented goals and intentions that guide their actions; and (ii) attention and knowledge as means by which they assess situations in order to pursue those internally represented goals and intentions effectively.

With respect to goals and intentions, even young infants identify the goal-object toward which an agent is acting (Woodward, 1998). But there is no evidence that infants understand the agent's internally represented goal or his intentions with respect to that goal, and some evidence that they do not. Behne et al. (2005) had an adult tempt infants and toddlers (at 6, 9, and 12 months) with an out-of-reach toy, which he never gave to them (in any condition). In several versions of an Unable condition, the adult did various things to show that his intention was to give the toy to the child, but he was unable to do so because of various obstacles (e.g., it would not fit through a hole). In several versions of an Unwilling condition, the adult performed actions similar to these but intended to keep the toy for himself (e.g., the hole was big enough, but he kept the toy anyway). The 6-month-old infants did not distinguish between the different intentions in the two sets of conditions. In contrast, the 9- and 12-month-old toddlers treated the conditions differently. When the adult was willing but unable to transfer the toy, they tended to wait patiently, presumably because they understood that his intention was to give them the toy. But when the adult performed similar actions that expressed an unwillingness to transfer the toy, toddlers showed impatience with his selfish intentions and requested the toy insistently, presumably because they believed that his intention was to keep it for himself (see Call et al., 2004, for similar findings with great apes in this so-called unwilling-unable paradigm). The toddlers—but not the infants—thus distinguished between similar actions that were underlain by different goals and intentions.

Other experimental paradigms reveal toddlers' intention understanding as they attempt to imitate others' actions. Imitation is an especially revealing response measure because children actually act out what they understand the other person to be doing. In his classic study, Meltzoff (1995) showed 18-month-olds a person attempting but failing to reach a goal. The toddlers did not reproduce the failed action, but rather the action they assumed the person was intending to produce (see Bellagamba & Tomasello, 1999, for the same finding with 12-month-olds, and Tomasello & Carpenter, 2005, for the same finding with human-raised great apes). Along these same lines, Carpenter et al. (1998) found that when 14- to 18-month-olds saw a person performing in quick succession two actions on an object, one accidental ("Woops!") and the other intentional ("There!"), toddlers reproduced the intentional action only (no matter the order in which they occurred; see Tomasello & Carpenter, 2005, for the same finding with human-raised great apes). In all these studies, toddlers (infants below nine months have not been tested) reproduced not the exact action of an agent but rather the action he intended to produce.

With respect to attention and knowledge, toddlers again show understanding beginning around the first birthday. Thus, it is at this age that toddlers begin to follow the gaze direction of an adult to a hidden target behind a barrier to see what she sees (Moll & Tomasello, 2004; see Tomasello et al., 1998, for this finding with great apes). Independent of gaze direction, toddlers of this same age also discern when another person knows something, in the sense that she is already familiar with it based on having perceived or attended to it earlier. In a study of 12- and 18-month-olds, Tomasello and Haberl (2003) had an adult approach a row of four objects, look generally at the row, and exclaim excitedly "Wow! Cool! Look at that! Can you give it to me?" The trick was that the toddler had previously shared attention to three of the objects with that adult, and so they were known to them, but they had never before shared attention to the fourth object (with that adult). Since it would be bizarre for the adult to express excitement about an object she already knew about, she was presumably excited about the new, unknown object, and indeed infants of both ages made this inference (see Hare et al., 2001, for a similar finding in a different paradigm with great apes). Beyond the studies of gaze following, in this study toddlers understood not just what the adult saw but what she knew.

Perhaps of most direct relevance to the hypothesis of common coding is the study of Meltzoff and Brooks (2008). They presented 12-month-old toddlers with an adult, who was wearing a blindfold, conspicuously orienting his head in a particular direction. As expected, most toddlers spontaneously looked in that direction (because they knew nothing of blindfolds). But if they were first given experience wearing the blindfold themselves—thus experiencing its blocking effects on their own visual access—toddlers looked much less often in the same direction as the blindfolded adult. In a second study, 18-month-olds were given experience with either a regular blindfold or a trick blindfold. Both looked opaque from the outside, but the trick blindfold enabled the wearer to nevertheless see through. In this case, after experience with both types of blindfold, the toddlers followed the gaze direction of an adult wearing a transparent (trick) blindfold but not one wearing an opaque blindfold (see Karg et al., 2015, for a similar finding with great apes). These studies, more than any others, show that toddlers are able to use their own visual experience to make inferences about the visual experience of others.

The conclusion is thus that toddlers first begin to understand others as intentional agents who operate with goals, intentions, attention, and knowledge in the same general age range that they themselves begin to operate in this way, namely, 9 to 12 months of age. Moreover, there is evidence that in at least some circumstances they can simulate the experience of other persons on analogy with their own experience. This new understanding of others, and simulation of others in intentional/mental terms, provides the raw material with which toddlers begin to construct their theories about why people do what they do.

4.3.2. Understanding Others' Decision-Making

Beyond understanding others' goals, intentions, attention, and knowledge, toddlers also understand aspects of how these work together in the decision-making process itself. Gergely et al. (2002) showed 14-month-olds (see Schwier et al., 2006, for 12-month-olds) a person performing a highly unusual action, namely, turning on a lamp with her head. In one condition, she did this in a straightforward manner with no constraints. In the other condition, her hands were occupied keeping a blanket wrapped around her shoulders, which could be seen as a constraint forcing the adult to use the unusual means of turning on the light with her head. When it was toddlers' turn with the light, they behaved differently in the two conditions. They took into account whether the demonstrator performed the unusual action as a free choice given no constraints—in which case they reproduced it—or else as forced by the external constraint of occupied hands—in which case the toddlers used a different means, such as using their hands to turn on the light, since they did not have those same constraints themselves. (Buttelmann et al. [2007] found something very similar with human-raised great apes.) Toddlers in this experiment were assessing *why* the adult chose to act as she did—*why* she chose this unusual action plan—which required them to think about the actor's alternative plans or intentions, that is, what else she could have decided to do given the constraints in the situation.

Also to investigate toddlers' understanding of decision-making, Moll and Tomasello (2006) had an adult enter a room with a basket searching for his lost toy. The 24-month-old toddler sat across from him and could see two toys, one out in the open and one on her own side of a barrier. The adult looked in the direction of each toy (i.e., for the one behind the barrier he looked at the barrier) and then repeated that he needed his toy. Obviously, he could not be talking about the toy out in the open that he could see or else he would have just taken it. The toddlers handed him the toy from their side of the barrier, the one he could not see, presumably understanding that his decision to keep searching for the toy instead of just grabbing the one out in the open was because the one he wanted was not visible to him. (See Hare et al., 2000, for a similar finding with great apes.) The toddlers understood how the adult's attention and knowledge affected her decision-making.

Importantly, again with respect to decision-making there is evidence that in at least some situations toddlers understand the intentional actions of others on analogy with their own intentional actions. Sommerville et al. (2008) performed a training/intervention study similar to the "sticky mittens" training studies with infants but now using tools. They demonstrated for some naïve 10-month-olds how to use a cane as a tool, and the toddlers learned how to use the new tool planfully. After this training, the toddlers who successfully learned how to use the new tool planfully then understood the planful (intentional) actions of a new adult in a way that toddlers who did not receive this training did not. Importantly, control subjects observing others using canes as tools did not show these same effects. This training study thus provides support for the hypothesis that children can simulate the intentional actions of others on analogy to their own intentional actions. [Once again, the strong version of the hypothesis—that children can only understand specific intentional actions of others if they have performed those same actions themselves—is likely too strong. A more moderate hypothesis is that children understand the intentional states of others in the same terms that they understand their own, and performing a particular intentional action themselves sometimes facilitates attribution of the corresponding intentional states to others by drawing attention to some special features of the situation.]

As in the case of causality, in the case of intentionality toddlers' inferences and predictions do not occur as isolated judgments. Rather, toddlers (and apes) make productive inferences organized into logical paradigms, in this case what is sometimes called practical reasoning based on the logic of agentive action (i.e., the practical syllogism). Thus, for example, toddlers and apes understand the intentions and knowledge of an agent enough to predict that he will choose to go for one object rather than another if and only if (i) he has the goal/intention/desire of having the object, and (ii) he perceives/knows/believes its location (e.g., at location A, not blocked by an occluder). The proto-conditional inference is thus: if that agent wants the reward and sees it at a location, then she will go for it there. They also make inferences based on proto-negation in terms of polar opposites: if the person sees only the occluder (not the reward), then she will stay (not go for it). And they can make exclusion inferences in paradigms similar to those they make about the physical world. In the rational imitation paradigm of Gergely et al. (2002), toddlers and apes understand that the physical constraints of the situation are forcing the agent to decide for an unusual action. The most natural interpretation is that the toddlers are employing a disjunctive syllogism from effect to cause with proto-negation: (i) if he had a free choice, he would be using his hands; (ii) he is using his head; (iii) therefore his choice must be externally forced (not free).

Toddlers and apes thus employ both a kind of proto-conditional and a kind of proto-negation to construct proto-logical (practical) paradigms licensing novel inferences and predictions about the intentional actions and decision-making of other persons. Once again, the hypothesis is that at the root of this all is the organization of the child's own agentive actions and the attribution of that organization to others based on common coding. Toddlers make decisions by comparing the current situation to the goal-situation and then formulating appropriate action plans, and so that is how they understand the intentional actions of others whom they observe and sometimes attempt to learn from, forming logically organized theories about why individuals are doing what they are doing in particular circumstances.

4.3.3. The "Simulation + Theory" Theory

There is thus good evidence that just as young infants begin producing goal-directed actions and understanding them in others at around the same age (~5 months), toddlers begin producing intentional actions and understanding them in others at around the same age (~9 to 12 months). The hypothesis is that these synchronies are a direct result of humans' (and other great apes') evolved system for the common coding of self's and others' actions (perhaps evolved for aligning intentional states with others in social learning and imitation).

This system of common coding enables children to understand the actions and experiences of self and other in the same basic terms, which for intentional agents comprise such things as intentionally structured plans, attention, and knowledge. But in a particular situation another person may have different plans or knowledge than the toddler, and so the toddler must form a theory or hypothesis to understand that person's actions in that situation. This can be seen, for example, in the Gergely et al. (2002) study in which the toddler understands the actions of the demonstrator in intentional terms but at the same time discerns that, because of the constraints of the situation, the demonstrator must act in ways that are not relevant for the toddler herself (who does not have those same constraints). The theories toddlers construct about why people do what they do thus comprise two sets of processes that are often pitted against one another in the literature: (i) simulation of the intentional states of others on analogy to one's own, which, in the current account, provides the requisite concepts; and (ii) theorizing about why others do what they do in a particular situation based on their particular intentional states in the circumstance at hand. We might thus call this composite account "the simulation + theory" theory.

Abstracting and synthesizing across multiple observations via processes of constructive thinking and re-representation on the executive tier leads toddlers to ever more powerful hypotheses and theories about how and why others do what they do. Nevertheless, theories based on observing the behavior of others are not the whole story. Toddlers also learn important but different things about others by actively participating with them in joint agencies to make joint decisions to pursue joint goals with joint attention. To make these joint agencies work, toddlers must actively coordinate intentional states with a partner—often using one or another form of cooperative communication—which requires such things as understanding the partner's role and perspective and making socially recursive inferences about them. The discussion of toddlers' learning and re-representation in the social domain is thus to be continued in the next chapter.

4.4. Toddlers as Intentional Agents and Theorists

The earliest mammals on planet earth had to compete with groupmates for resources, creating extra challenges for making effective and efficient behavioral decisions. They met these challenges by evolving a psychological architecture of intentional agency that enabled them to cognitively simulate possible actions and their likely results—that is, to think and to plan—before actually acting, as well as to actively imagine what might possibly happen in the environment. This all required imaginative cognitive representations that individuals could evoke voluntarily.

The hypothesis I have explored in this chapter is that human toddlers, from nine months to three years of age, operate in this same general manner. Again, one may ask why it is adaptive for human toddler psychology to be organized in this way, that is, beyond that of infants. We can only speculate, but a reasonable hypothesis, as noted earlier, is that 9–12 months of age is when infants turn into toddlers in the literal sense that they begin to locomote more independently, including learning to walk. Locomoting more independently means making more agentive decisions on one's own, and indeed there is a large body of research showing that the emergence of self-produced locomotion in toddlers is accompanied by a whole host of new capacities, including "social and emotional development, referential gestural communication, wariness of heights, the perception of self-motion, distance perception, spatial search, and spatial coding strategies" (Campos et al., 2000, p. 149). With the advent of self-produced locomotion, toddlers enter a new world of challenges for effective and efficient decision-making.

The evidence reviewed in this chapter suggests that human toddlers do indeed operate as individual intentional agents. With their new manner of agentive functioning, toddlers are able to embed one action plan within another to form hierarchically organized action plans, or intentions, to do such things as remove obstacles so as to gain access to opportunities. This kind of action planning involves cognitive simulations of potential actions and their likely results in the environment, among which the infant chooses in acts of either/or decision-making (involving proactive inhibitory control of the unchosen option). None of this can be done with young infants' iconic representations tied to perception; it requires the voluntary evocation of representations and their imaginative manipulation and suppression. And most importantly, all of these new types of cognitive functioning require a fundamentally new psychological architecture, that is, an architecture comprising not just agentive decision-making and action, but also an executive tier of proactive monitoring and control. This executive tier of operation provides a workspace (executive working memory) for the use of imaginative representations in thinking, planning, and the compiling of hierarchically structured actions.

This general mammalian way of acting in the world intentionally is supplemented in human toddlers by two new ways of understanding the world—two new dimensions of experience—as inherited from their great ape ancestors. First, the earliest great apes evolved a capacity for making and using tools, which provided a bridge for understanding causal relations among external events. Toddlers develop this capacity soon after nine months of age as they start to incorporate tools into their intentional actions, which they then leverage into an understanding of entire logical paradigms of causal inferences (derived from the logic of action). Second, based on great apes' evolved capacity for the common coding of the psychological processes underlying the actions of self and other—very likely evolved to facilitate apes' aligning of their own intentions and actions with those of others in social learning—toddlers become able to understand others' intentional actions, also structured into logical paradigms of intentional inferences. These newly constructed causal and intentional paradigms of logical inference provide the basis for children's first theories of the world, as they for the first time create the possibility of understanding *why* physical events occur in the ways they do and why other people act in the ways they do. Interestingly, these two grand explanatory schemas—sometimes referred to as mechanistic (or causal) versus functional (or intentional) schemas—continue to structure children's theories into adulthood, and throughout ontogeny humans display a tendency toward animism or "promiscuous teleology" in which they sometimes explain physical events in intentional terms (see, e.g., Kelemen, 2004). In the other direction, some neuroscientists hold out the hope that we will someday be able to explain intentional states in mechanistic terms.

The understanding of causality and intentionality also generate new forms of learning. As toddlers act on the world, they are able to engage in a kind of means-ends analysis that leads to assessments of how their actions cause (or do not cause) various effects in the external world, which leads to much more efficient and effective learning. Executive regulation of this kind of learning leads to actions specifically designed to test hypotheses about specific causal relations in the world, what I have called hypothesis-directed learning. Moreover, toddlers' social learning focuses only on the intentional (not accidental) actions of others, which leads to more efficient and effective social learning as well. Understanding causality and intentionality thus enables toddlers to formulate and test hypotheses and theories about why things happen as they do, empowering them to learn many new kinds of things in many new ways. And all these new competencies emanate, more or less directly, from the way the child as agent makes things happen with her actions (including with intermediaries) and executively regulates the process.

In general, operating in all these new ways requires that toddlers navigate not just a world of actualities but also a world of **possibilities**: actions may be imagined before execution, and causal hypotheses may be imagined before testing. Perhaps surprisingly, this way of operating is not unique to humans; in a general way, it is also characteristic of other great apes. But toddlers are not simply great apes. On the one hand, they are much less competent than other apes in their behavioral decision-making, as can be seen in their inferiority to apes in such things as tool-making and innovation, finding their way in large spaces, and complex planning in spatial mazes (see Call & Tomasello, in press, for a review). Also, as we shall see in Chapter 6, toddlers have no skills of metacognition as do great apes (and preschoolers). But, on the other hand, toddlers possess a suite of social-cognitive skills that great apes do not possess. These are based on toddlers' ability to form with others as joint agencies, which involves not just attributing intentionality and causal efficacy to others, as great apes already do, but on an ability to mentally coordinate with others and to jointly self-regulate this coordination—as we shall now see.

5
Joint Agency and Perspectival Representations

While toddlers are becoming more competent intentional agents who think, plan, and executively regulate their actions, they are also beginning to engage with adults in some new and unique forms of collaboration and communication. These new forms of collaborative and communicative engagement are presaged by infants' emotion sharing with caregivers in protoconversations. But then at around nine months of age, as just reviewed, toddlers become intentional agents themselves and begin to understand others as intentional agents as well. These new competencies, along with newly emerging cooperative capacities and motivations, enable toddlers to form with others joint agencies.

In joint agencies, toddlers coordinate their intentional actions with other intentional agents in species-unique ways by creating both joint goals and joint attention. The social-cognitive capacities empowering the creation of joint goals and attention are what we may call joint intentionality. Other primate species do not create joint agencies underlain by joint intentionality because joint goals and attention present serious coordination problems (in the game theory sense). These can only be solved by a shared (common ground) understanding of what "we" are doing or attending to, which requires the ability to mentally coordinate recursively. To facilitate this coordination, toddlers develop, in addition, species-unique forms of cooperative-referential communication, beginning with the simple but powerful pointing gesture, followed by symbolic (including linguistic) communication. The evolutionary hypothesis is that only humans evolved to *co*-operate in all these ways because only they evolved in the face of ecological challenges requiring collaborative activities structured by joint intentionality (see Chapter 2).

The modern theory of cognitive development recognizes and investigates young children's collaboration, joint attention, and communication. But it does not appreciate sufficiently, in my view, the novel cognitive representations and processes that make these unique interactions possible. That is, it does not recognize that uniquely human collaboration and communication require a joint agency comprising: (i) socially shared goals and attention, which create a kind of shared reality; (ii) individual roles and perspectives within these shared realities; and (iii) recursive mental coordination of perspectives via cooperative-referential communication to facilitate the joint agency's functioning. Spelke (2022) attempts to capture all of these phenomena by positing that young toddlers begin to understand that cognitive representations are "shareable." But this just labels the phenomena without explaining them. To explain joint agency and joint intentionality we must posit some new cognitive representations and operations.

5.1. Collaboration and Joint Intentionality

Uniquely human collaborative activities first emerge in ontogeny at around nine months of age as toddlers collaborate with adults in activities such as building a block tower together, putting away toys together, and getting the child dressed together. Longitudinal studies with standardized assessments confirm the mean and modal age of emergence of such activities as nine months (Carpenter et al., 1998), and cross-cultural assessments have found common ages of emergence for these activities across diverse cultural contexts (Callaghan et al., 2011). Epistemically, the partners in a joint agency engage with one another via joint attention to create a common ground understanding of the situation at hand, a kind of ad hoc shared reality.

5.1.1. Joint Goals and Joint Attention

Forming a joint agency to pursue a joint goal constitutes a coordination problem: I want to build the block tower together with you only if you want to build it together with me, and you are thinking in the same way in reverse. This can lead to paralysis unless the two of us know together in common ground that we both want to build the tower together (Bratman, 2014). This knowing together in common ground constitutes a kind of recursive engagement in which each partner knows that the other knows that I know ... that we both want to build the tower together (a recursivity well-known since the pioneering work of Schelling, 1960, and Lewis, 1969). Forming joint agencies in this way is unique to the human species, as many studies show that although great apes act together in some ways, they are not bound by a recursively structured joint goal, whereas human toddlers are acting together with their partner as a recursively structured joint agent "we" (see Tomasello, 2022c, for a review of empirical studies supporting this claim).

Acting together collaboratively requires joint attention. Again, this creates a coordination problem, and again the solution is the recursive coordination of attention to create a common ground understanding. Each partner attends to the other attending not only to a situation but also to her attention to that situation, as well as to her attention to their attention to the situation, etc., which constitutes a common ground understanding that "we" are jointly attending to the situation (Tomasello, 1995; 2014). For example, in a recent comparative experiment an ape or human toddler entered a room to find a human adult (back turned) looking at a TV screen, which then began showing a film. In this control condition, the subject knew that they were both watching the film but knew nothing of the adult's knowledge of her and her presence. In the experimental condition, however, as soon as the film began the adult turned and looked briefly to the subject—before returning to the film. In this case, the subject could potentially know that we both know together that we are jointly watching the film. (NB: In the control condition the adult also looked to the child, but only later in the procedure.) The apes did not differentiate these conditions, whereas the toddlers formed a closer social bond (approached more readily) with the adult, who jointly attended to the film with them (Wolf & Tomasello, 2020). The natural interpretation is that in the experimental condition the toddlers, but not the apes, created with the adult a common ground understanding that they were watching the film together in joint attention.

Importantly, toddlers form joint agencies almost exclusively with adults, and indeed toddlers' attempts to collaborate with same-age peers are very uncoordinated until 2.5 to 3 years of age (Brownell & Carriger, 1990). In terms of joint attention, toddlers do not seem to share attention with same-age peers in anything like the way they do with adults, again until something like 2.5 to 3 years of age (see Tomasello, 2020a, for a review). Toddlers are simply not adapted for collaborating with peers, but only with adults. In such asymmetrical collaborative activities, the more competent adult does whatever it takes to make sure that each partner plays their role and attends to relevant things: she regulates the process—making or structuring decision-making for the joint agency—with some sophisticated metacognitive skills that the toddler does not possess. And so even when collaborating with an adult, toddlers are only good at coordinating with a partner in terms of concrete actions and attention, not in terms of coordinated thinking, planning, and decision-making (since this would require a second-order metacognitive tier of functioning).

And so the overall picture is that even though great apes operate with a second-order metacognitive tier of individual functioning (see next chapter), they are not adapted for joint agency. In contrast, human toddlers are adapted for joint agency, but only with more competent adults, who scaffold the interactive process in ways that compensate for toddlers' inability to metacognitively regulate their executive-tier processes. Toddlers already employ many uniquely human cognitive representations, operations, and skills, but at the same time their skills of agentive action (both individual and joint) are limited without adult support.

5.1.2. The Importance of Roles

Joint goals create individual roles; without a joint goal we are just doing different things. This means that joint agency comprises a kind of dual-level structure: on the higher level is the joint goal (e.g., "we" are building together a block tower) and subordinate to this are the individual roles of the partners (e.g., "you" are holding the foundation steady while "I" am placing a block on top). Comparative studies show that toddlers understand the way that joint goals create individual roles in a way that great apes do not. For example, toddlers (i) simulate a partner's role in a way that apes do not, (ii) engage in a reversal of roles in a way that apes do not, and (iii) coordinate partner roles via acts of cooperative communication in a way that apes do not (see Tomasello, 2022c, for a review of these and other studies). Toddlers' understanding of collaboration via joint agency, then, includes an understanding not only of the joint goal but also of the individual roles involved.

Toddlers' understanding of roles in joint collaborative activities has two momentous consequences for their cognitive development. First, it creates a new relational dimension of human thinking unavailable to other ape species. It is not that great apes do not understand relations at all; they understand various spatial and quantitative relations of magnitude such as bigger-smaller, brighter-darker, fewer-greater, higher-lower, and some forms of same-different (see Call & Tomasello, in press, for a review). But human children understand a much wider variety of different types of relations (Gentner, 2003). In particular, they understand, in a way that other apes do not, functional categories of things defined by their role in a larger activity, what we may call the relational-thematic-narrative dimension of human cognition. Thus, humans are exceptional in creating categories such as *pet, teacher, sister, doctor, hider-seeker, guest, father, lawyer,* and so forth, what Markman and Stilwell (2001) call "role-governed categories." These are relational not in the sense of relating things on a physical dimension, but rather in specifying the dynamic relation between an entity and some larger event or process in which it plays a role. My hypothesis is that this species-unique relational-thematic-narrative dimension of human experience and thinking derives directly from humans' understanding of collaborative activities in terms of shared goals and individual roles. The grammatical constructions of a human language symbolize many different ways of organizing experience via this dimension (see Section 5.3.2).

Second, humans' unique understanding of the roles constituting joint intentional activities provides an initial step into normative attitudes and thinking. These are also, as far as we know, unique to the human species, and they will become of special importance after three years of age. The idea is that in a joint collaborative activity the partners know together in common ground the ideal way in which each role should be played: the person holding the foundation block should hold it steady, and the person placing the block on top should do so gently and carefully. Failure to live up to these mutually known ideal roles will lead to joint failure and so to partners being upset with one another or feeling bad for letting the other down. My contention is that it is only when individuals are acting together with one another as a joint agent—when each depends on the other to play her role appropriately for joint success—does this normative dimension of human experience arise (Tomasello, 2020b). How this develops into the normative concepts and attitudes of children after three years of age is a major topic of Chapter 7.

5.1.3. The Importance of Perspectives

Joint attention creates individual perspectives; without joint attention we are just seeing different things. Joint attention is thus again dual-level: "we" are attending together to a block tower, with "you" attending to one side and "I" to the other. Great apes understand that others attend to things, but because they do not engage in joint attention, they do not understand the notion of different perspectives on the same thing. Toddlers do. For example, in the Moll and Tomasello (2006) study cited in the previous chapter, 24-month-olds understood that they and the adult had different perspectives on the situation: they themselves could see two objects whereas the adult could see only one (the other lying behind a barrier from the adult's viewing angle). The toddlers in this study were engaging in what is known as level-1 perspective-taking (Flavell, 1992): they knew that what they and the adult saw in the situation was different. Level-1 perspectives are best thought of as "attentional perspectives," whereas the level-2 perspectives characteristic of older children are best thought of as "conceptual perspectives": seeing one and the same thing as different under different conceptual descriptions (e.g., as dog or animal). Taking the perspective of another person, of whichever type, is at bottom an act of imagination, which toddlers are capable of doing via their imaginative representations.

Toddlers' understanding of attentional perspectives in joint attention has two momentous consequences for their cognitive development. First, operating within a shared reality in which each participant understands that their partner nevertheless has a distinct perspective on the situation is the interactive framework within which toddlers learn and use their unique forms of cooperative/referential communication, both gestural and linguistic. Engaging with partners in joint intentional collaboration and cooperative/referential communication are prerequisites for toddlers to construct perspectival cognitive representations, that is, representations that go beyond the basic imaginative representations of toddlers as individual agents by contrasting different ways that partners in joint attention may attend to a situation. How this works in both gestural and linguistic communication during the toddler years is the topic of the next two sections of this chapter.

Second, in addition, operating within a shared reality in which each participant understands that their partner nevertheless has her own perspective on the situation constitutes a key first step in understanding the distinction between individual perspectives or beliefs, on the one hand, and the objective situation as independent of all perspectives and beliefs, on the other. The distinction between subjective perspectives and the objective situation underlies a variety of uniquely human concepts and constitutes one of the major cognitive-developmental achievements of children after three years of age. How children scale up from the notion of attentional perspective to the notion of conceptual perspective and beliefs, sometime after three years of age, is a major topic of Chapter 7.

5.1.4. Self-Other Equivalence

Finally, participation in joint agencies during the one- to three-year age period leads young children to still one further fundamental insight: my partner's individual goals and perspectives are just like mine. That is, through their participation with partners in joint agencies, young children cannot help but recognize that (i) our respective role contributions are equal causal forces in realizing our joint goal, (ii) we could reverse roles and still have an intact joint agency, and (iii) the ideal way the role must be played is the same for us both impartially, independent of our particular characteristics. The outcome is that by the end of toddlerhood, children recognize that their partner's goals and preferences, as well as their partner's perspectives, are, in all essential respects, equivalent to their own.

Recognition of the equivalence of self and other in a joint agency is of special importance as it is the foundation of uniquely human objective and normative thinking. Epistemically, the recognition that my partner's perspective is different from mine but potentially just as valid is the *sine qua non* for constructing the notion of an objective (impartial) perspective on things. And it provides the context within which partners can have a reasoned dialogue in which each respects the perspective of the other as they try to find a common ground understanding to guide them in deciding what they should believe or do. Morally, the recognition that my partner' goals and preferences are different from mine but potentially just as valid is the *sine qua non* for constructing the notion of fairness—based on judgments in which everyone, including the self, is treated equivalently and impartially—as well as social norms in which all individuals subject to the norm are treated equivalently and impartially. And again it provides the context within which partners may have a reasoned dialogue in which each respects the values of the other as they try to find some common ground values to guide them in deciding what they should believe or do. How this recognition of self-other equivalence contributes to the development of objective/normative representations and concepts, sometime after three years of age, is a major topic of Chapter 7.

5.1.5. Cognition for Collaboration

Collaboration in joint agencies thus requires toddlers to understand:

- joint goals, joint attention, and common ground;
- individual roles, individual perspectives, and self-other equivalence; and
- the recursive structuring of goal and attentional coordination between collaborative partners.

My hypothesis is that the development of these species-typical capacities into psychological realities requires actual participation with others in joint agencies. Whereas modern cognitive-developmental theory focuses almost exclusively on children as third-party observers, the current agency-based model posits that active participation in joint agencies provides a special context for cognitive development because it forces the child to deal with a partner with whom she shares goals and attention but who has his own role and perspective. In direct engagement with a partner, it is especially clear that one role presupposes another and one perspective presupposes another. Without a simultaneously interacting partner, the notion of a solitary role or perspective is just, so to speak, the sound of one hand clapping.

5.2. Cooperative/Referential Communication

Humans communicate with one another in a number of unique ways, from the pointing gesture to language. Almost certainly the original evolutionary function of these unique forms of communication was the coordination of roles and perspectives in collaborative activities (Tomasello, 2008). And indeed, human communication is itself a collaborative activity. We have the joint goal that you comprehend my communicative act. I play my role by trying to formulate my act perspicuously for your perspective, and you play your role by making good faith attempts to comprehend my communicative act, asking for clarification as needed. The communication of other species is not cooperatively structured in this same way.

Given our developmental model, human children should not begin engaging in this kind of cooperative communication—involving roles and perspectives—until after nine months of age. And this would seem to be the case. Although young infants can spur adults into action with various ritualized acts, only after the 9–12-month transition does cooperatively structured communication emerge, mainly via the pointing gesture. The mean and modal age at which young toddlers in western cultures begin to point is 11 months (Rüther & Liszkowski, 2023), which is the same age at which they show comprehension of pointing (Carpenter et al., 1998). In a large elicitation study across seven diverse societies, all of the almost 100 toddlers first pointed between 10 and 14 months of age (Liszkowski et al., 2012), and a study that attempted to train western infants to point earlier than normal failed to produce accelerated competence (Matthews et al., 2012).

The traditional view is that toddlers use the pointing gesture with two communicative intentions: imperative (e.g., to request an object) and declarative (e.g., to draw attention to an object) (Bates, 1979). But they also use the pointing gesture informatively to apprise someone of something he does not know, so-called informative pointing, which has been experimentally demonstrated in 12-month-old toddlers. For example, when an object falls off a table silently onto the floor, if the adult sitting at the table clearly witnessed this event, the 12-month-old toddler does nothing. But if the adult is looking away at the time, when he looks back to the toddler she immediately informs him of the object's new location via a pointing gesture (Liszkowski et al., 2008). I focus here on informative pointing of this type because it is especially interesting from a cognitive point of view.

5.2.1. The Referential Intention

Human cooperative/referential communication is indirect; it comprises at least two intentional levels. The communicator invites the recipient to jointly attend with her to some situation (the referential intention), but she is doing this only because she has some larger motive that the recipient know or do something (the social intention). Thus, I want you to attend with me to the box over there (referential) because I want you to know where the missing toy is located (social).

The process of establishing joint attention on some referential situation is not a one-shot, ballistically produced intentional action, but rather a process of cooperative coordination. That is to say, one individual indicating and another individual identifying the intended referent of a pointing gesture involves the coordination of attentional perspectives. In the prototypical case, one partner initiates things by pointing for the other to a referent that she (the communicator) is already attending to; her (referential) intention is the aligning of their attention in joint attention. The recipient, if he is being cooperative, goes from his own individual attention elsewhere to jointly attending with the communicator. The interpersonal coordination thus involves each partner's sequential shifting from individual to joint attention, as either communicator or recipient, with adjustments as needed (Liszkowski et al., 2007). Unlike simply imagining what another person sees or knows, as occurs in many studies of infant social cognition, negotiating joint attention brings into focus the *relation* between self and other perspectives. To know that perspectives are or are not aligned there must be some imagining of the content of those perspectives and their relationship. Such negotiations require both imaginative representations and an executive workspace in which the two attentional perspectives may be imaginatively compared and coordinated.

From as soon as they begin communicating intentionally, young toddlers can use informative pointing to refer to situations much more complex than an object to which they wish to draw attention. For example, in a multi-child diary study, Carpenter et al. (in preparation) report the following examples from children who have either not begun talking at all or who use only a few simple words (the glosses for these are generally consistent with the adult's immediate interpretation):

> Example 1: At age 11.5 months, J points for Mom to the door as Dad is preparing to leave. GLOSS: I want you to know that Dad is soon leaving out that door.
> Example 2: At age 13 months, J watches as Dad arranges the Christmas tree; when Grandpa enters the room, J points to the tree for him and vocalizes. GLOSS: I want you to know that there are new things on the Christmas tree.
> Example 3: At age 13.5 months, Mom is looking for a missing refrigerator magnet, and L points to a basket of fruit where it is (hidden under the fruit). GLOSS: I want you to know that the magnet you are seeking is there.
> Example 4: At age 14 months, two different children, J and L, have accidents when a parent is not looking. When the parent comes to investigate, the infant points to the offending object (i.e., the thing he bumped his head on, or the thing that fell down). GLOSS: I want you to know that this is what hurt me or fell over.

What stands out in these examples is that the young toddlers, even before they have much or any language, are attempting to communicate some fairly complex facts about the world. In general, children's informative pointing (just like their later informative language) indicates an entire situation in a quasi-propositional manner. Thus, in these examples they are informing an adult of the fact that Dad is soon leaving out that door, or the fact that there are new things on the Christmas tree, or the fact that this is the object on which I bumped my head. In cases in which they seem to simply be pointing out an object, they are actually indicating something more propositional, like the fact of the object's presence or location. These referential intentions are not explicitly propositional because they do not make explicit the elements and relations involved; but they are implicitly propositional in that these elements and relations are a part of the referential intention in a kind of holistic undifferentiated manner. The complexity of the referential situations about which young toddlers are capable of communicating contrasts sharply with the "pointing" of chimpanzees for humans (e.g., Leavens & Hopkins, 1998), which is almost always a simple request for an object—perhaps expressing a kind of ritualized reaching.

Recognizing the powers of conceptualization at work here is important because most studies of early cognitive development focus on infants' and toddlers' understanding of such things as objects, space, number, causality, and other general dimensions of experience, positing that the understanding of propositional structures emerges only with grammatical language. But here we see that toddlers as young as 12 months of age can understand and indicate—at least holistically—complex, quasi-propositional situations, and they do this well before they produce grammatically structured language. It is of course possible to posit that children comprehend grammatically structured language already at this age, and their pointing to quasi-propositional referents is somehow derivative of this ability. But the hypothesis here is that it is the other way around: attention and joint attention are from the beginning focused on whole situations, with informative pointing conceptualizing such referential situations in a quasi-propositional format for a recipient, and grammatical language then symbolizing them in a fully propositional format (see Chapter 8 for the argument that this had to be the case in human evolution and history).

5.2.2. Embedding the Referential Intention in the Social Intention

Toddlers also comprehend the pointing gesture in a way that neither young infants nor great apes do. The reason is that to understand what someone is pointing to (his referential intention) one must first understand why he is pointing to it in the first place (his social intention). Thus, if an adult points for a toddler in the direction of a toy across the room, he could be referring the toddler's attention to any one of an infinite number of situations: the location of the toy (if they are looking for it), the color of the toy (if they are searching for pink things), the brokenness of the toy (if the adult wants to point out why they cannot use it), or whatever. The child must understand why the adult assumes that pointing in a particular direction is *relevant* for her in the current situation (Sperber & Wilson, 1986). The search for relevance is prompted by the toddler's natural recognition that the adult's communicative act is cooperatively "for me," as expressed in such things as eye contact, calling the child's name, etc. Without this recognition, the child simply sees the adult's act as a protruding finger that has no relevance for her (Schulze & Tomasello, 2015).

Consider the following simple experiment. An adult engages the child (or ape) in a hiding-finding game in which it is made clear that the child's role is to find the toy and the adult will help her do this in various ways. In the key trials, the child watches as the adult hides the toy under one of two cups, but this takes place behind an occluder so the child does not know under which cup. Then the adult points to one of the cups to indicate the hidden object's location. In this case, infants younger than nine months (and nonhuman great apes) choose between the cups randomly. They do not comprehend that the pointing gesture is intended as a communicative act to help them find the hidden object. But why not? They can follow the adult's gaze direction to one of the cups, and they know the object is under one of the cups. But still, they do not make the connection, as 12-month-old toddlers readily do (Behne et al. 2005; 2012). One hypothesis is that the toddlers, but not the infants or great apes, understand the adult's recursively structured intention: they understand that he is pointing to that cup because he *intends* for me to *know* where the toy is located. Interestingly, great apes and young infants do not understand such embedded communicative intentions even if the adult makes an iconic gesture (or holds up a replica) that resembles the cup under which a toy is hidden (Bohn et al., 2019; 2020). They fail even though other assessments show that they can see the resemblance between the iconic gesture and the object's hiding place. They do not understand that the communicator intends for me to infer his social intention by discerning his referential intention and its relevance to that social intention.

This all works only if the young toddler appreciates the common ground she has with a particular individual as established by previous joint attentional interactions. For example, if an adult and 14-month-old toddler are cleaning away toys into a basket, and the adult points to a toy next to the wall, the toddler toddles over, fetches it, and puts it in the basket with the others. But if a different adult enters the room with no knowledge of the joint cleaning up and points in the same fashion to the same toy, toddlers know that this new adult knows nothing of the toy as a target of cleaning up—so instead of cleaning it away in the basket, they simply hand it to the new adult (Liebal et al., 2009). Joint agency (cleaning up together) creates joint attention to relevant aspects of the situation, which creates the common ground necessary for the toddler to interpret the otherwise ambiguous pointing gesture appropriately. And, of course, bottom-up joint attention—for example, the child and the adult are jointly attracted to a singing bird—also creates a common ground that may be used to interpret a pointing gesture. Joint attention and common ground are crucial in enabling all kinds of complex social coordinations that would otherwise not be possible, all the way into adulthood (Clark,1996; Thomas et al., 2014).

The child also uses common ground in her productions. The interplay between what is in common ground—already given or shared—and what is new for the recipient (as best the communicator can determine it) is especially clear in the following example from Carpenter et al.'s diary study (in preparation). A 14-month-old toddler always wants to see his highchair in its place at the dining room table. On one occasion he and his mother enter the dining room and come upon the chair next to the window. The toddler then points to the table where he wants his chair to go. On another occasion, he and his mother enter the dining room and come upon the table first. Upon spying his chair across the room, he points to it, seemingly with the exact same communicative intention, namely, that it be put in its place at the table. The toddler in both cases is assuming as common ground the first object to which they jointly attend (either chair or table) and then points to the other key element, as new, to express what he wants (which Mom infers). It is noteworthy that again in this example the child clearly understands the complex situation—the relevant actions, objects, and locations in a quasi-propositional structure—in a manner that will later support a grammatically structured linguistic utterance (see next).

5.2.3. Cognition for Cooperative/Referential Communication

Toddlers thus employ in their comprehension and production of gestural communication some complex cognitive capacities that are not sufficiently appreciated by the modern theory of cognitive development with its almost exclusive focus on children as third-party observers of the world. Most importantly, cooperative/referential communication requires toddlers to:

- produce and comprehend recursively structured communicative intentions;
- understand the quasi-propositional content of communicative intentions; and
- coordinate attentional perspectives to achieve joint attention.

Once again, the hypothesis is that the development of these species-unique capacities into psychological realities requires the toddler to actually participate in cooperative/referential communication. Finding a way to get another individual to join with me in seeing a situation from a particular perspective requires a creative act of constructive thinking, as well as the ability to coordinate perspectives on the executive tier of operation. And something similar takes place in comprehension when the toddler attempts to abductively infer how a communicator wants her to perspectivize a situation. Negotiating perspectives in these ways leads toddlers to re-represent on the executive tier a new kind of pragmatic knowledge about how one's own attentional perspective relates to that of others in particular situations. Great apes neither participate in cooperative/referential communication nor show evidence for this kind of pragmatic knowledge in other contexts.

5.3. Linguistic Communication

Toddlers' skills of linguistic communication are continuous with their skills of cooperative/referential communication using the pointing gesture. Both rely on cooperation, joint attention, embedding referential intentions in social intentions, common ground, attentional perspectives, and recursive inferences. It is just that in linguistic communication this is done with socially learned symbolic means, originally created by others in the culture for various communicative purposes.

These symbolic means comprise words and grammatical constructions. According to Tomasello (2003), the process of learning to use a linguistic symbol—either word or grammatical construction—involves two basic processes: (i) intention reading (including joint attention) to determine the speaker's communicative intentions or meaning; and (ii) pattern finding to construct a representational schema that abstracts across the different uses of a word or construction as observed in mature users over time (via learning and re-representation). Toddlers start with the simplest and most frequent adult uses of a word or construction and then add more uses as they observe them, possibly synthesizing all of those for a particular word or grammatical construction into its "meaning." Here I first consider the process of learning words and then the process of learning grammatical constructions.

5.3.1. Linguistic Symbols as Perspectival Representations

Given its dependence on the same basic cognitive capacities as the pointing gesture, skills of linguistic communication should not emerge until sometime after nine months of age. In apparent contradiction to this prediction, some studies have found that when 6-month-old infants hear a familiar word (e.g., *bottle*) they reliably look more to a picture of a bottle over other pictures (Bergelson & Swingley, 2012). However, it is possible that this is not really language comprehension but rather something more like association or recognition: infants look to the object that they have reliably seen when they have previously heard that sound. This ability has also been found in some domestic dogs (and parrots) who reliably fetch objects associated with "words" they have heard (Kaminski et al., 2004). The prediction here is that children should not produce or comprehend linguistic symbols as acts of cooperative/referential communication until after nine months of age, as evidenced either by the appropriate production of referential language or accurate comprehension in tasks requiring the behavioral selection of a referent. These skills typically appear at around, or soon after, the first birthday (e.g., Bates, 1979).

The process of learning a word is often characterized as learning to "map" a word onto an object or a meaning (e.g., Bloom, 2002). But obviously children are not mapping a word onto a physical object, and there is no pre-existing thing in the world called a "meaning" onto which they could map a word. So perhaps they are mapping words onto concepts. But in the right circumstances children can comprehend novel words even though they do not have a pre-existing concept (see Tomasello, 2001, for a review). An alternative approach is to jettison the mapping metaphor (and association learning theory) in favor of a more agentive approach. The first step is to recognize that words are just sounds until they are used by an intentional agent in an act of communication. Is *gavagai* a word? Only if people use it as one. And so what children are learning is how adults use words. An apt analogy is learning to use a tool. One can say that children learn to associate scissors with cutting—or to map scissors onto cutting—but it is more apt to say that children learn how to use scissors by observing how adults use them. And so children learn to use a word by observing how—for what communicative purposes—adults use it (this is why the philosopher Wittgenstein, 1953, claimed that "meaning is use"). To say that a child has learned the meaning of the word "play" is to say that she has discerned people's communicative intentions in using this word. This is not a one-shot deal. Thus, a child might start by learning the word "play" only for activities with toys, and then later come to include uses such as playing a violin, or playing a part in a play, or playing favorites in the classroom. The child comprehends how people use a particular word on a particular occasion via intention reading (with joint attention), and then, over time, finds a pattern in people's use and so synthesizes these (via learning and re-representation) into a meaning or set of meanings for that word.

The way children understand an adult's use of a word in a particular context is almost miraculous. Textbooks often depict the process as similar to classroom learning: an adult draws the child's attention to an object and names it. This may happen occasionally with some kinds of words (e.g., object labels) in some cultures (i.e., those that engage in language pedagogy, which not all do). But in most cases, discerning how an adult is using a novel word requires the child to employ her most powerful skills of intention reading and joint attention. For example, consider the following experiment. The adult says to the child, "Let's go find the toma." The two of them then go to a row of buckets, and the adult extracts an object from the first one, frowns at it, and puts it back. Associative learning theory would suggest that the child should learn the word *toma* for this object as it is most closely associated in space and time with hearing the new word. But then the adult extracts an object from the second bucket, smiles excitedly, and places it in her lap (after this she says, "Let's see what else is in here," extracts an object from the third bucket, smiles at it, and returns it). Eighteen-month-old toddlers learn the word *toma* as designating the second object (Akhtar & Tomasello, 1996), which they have singled out based on knowledge of the fact that people search for an object until they find it, bypassing other objects as needed, at which point they are content. In another experiment, when an adult announces her intention to "Dax" Big Bird, toddlers learn the word *dax* for the next action the adult produces intentionally, even if she produces an accidental action in between—because, in this case, they know that one does not announce an intention to produce an accident (see Tomasello, 2001, for a review). And it is difficult to see how children might learn the appropriate uses of words such as *thanks* and *goodbye*, except via intention reading.

Once children have learned some pieces of language, they do not just produce each of them associatively whenever they encounter the appropriate referential situations. Young children produce language strategically as a social/communicative tool, and if they want to communicate effectively, they must formulate their utterances in a way that facilitates listener comprehension in the situation at hand. Thus, from around two years of age, toddlers are skillful at taking the perspective of their listener to choose the best word for that listener (i.e., in simple situations in which such perspective-taking is relatively easy; see Vasil, 2023, for a review). The best example is 2-year-olds' growing skills for using pronouns such as *it* appropriately, since pronouns are used precisely when the child can judge that her intended referent is already in her common ground with her listener in a salient way. Thus, from around two years of age, toddlers strategically learn to use the word *it*, rather than *ball*, when their listener has just referred to "the ball" or when the two of them are manipulating or looking at a ball right in front of them. How children learn to do this is not at all clear (there is very little relevant research). But at the very least, in observing an adult refer to an object as *it*, the child must discern that the adult is taking her perspective and judging that she is already focused on that object. Then the child, to use the word *it* appropriately herself, must do the same in reverse, that is, she must take the perspective of her partner and discern when an object is already in her attentional focus. Mapping or association learning clearly cannot handle the learning of pronouns at all; it is rather a process of learning through intention reading and perspective-taking (supplemented by processes of constructive thinking and re-representation to discern general patterns).

The outcome is what we may call perspectival cognitive representations. Toddlers already take the perspective of an adult when he uses a pointing gesture in an attempt to align their attention jointly on a referent. But language introduces a means for symbolizing perspectives on a situation. This occurs in two different ways. First is what we may call profiling. Consider the common situation in which an adult has an object, and the child wants it. At 24 months of age, the child investigated by Tomasello (1992) had 10 different ways of requesting the object, depending on various features of the situation. Thus, she could just name the object she wanted. Or if the adult was keeping it away from her, she could implore "get it" as she tried to get it or "got it" after having done so. She could say "have it" if she was stressing simply that it be in her possession, or "give it" if she was requesting that the adult actually hand it over. She could say "hold it" for certain precious objects, like a kitten, that she wanted to actually hold in her hands. She could say "back" when she had just given the object to the adult previously and wanted it back, and she could say "another one" or "more" if the adult had just previously given her an object or substance of that same type. And finally, she could ask the adult to "share" the object with her (play with it together) or let her borrow or "use" the object (and soon return it). Such representations may be considered perspectival in the sense that the physical situation of an adult possessing an object right in front of the child is more or less identical in all of these various cases. I am calling this profiling because the referential situations are slightly different in the various cases if one takes into account what the child is focused on in her understanding of the situation. Figure 5.1 provides iconic representations of these different perspectival representations, with iconic representations (not words) again being most appropriate for depicting the child's visual attention that is being abstracted and perspectivized.

88   AGENCY AND COGNITIVE DEVELOPMENT

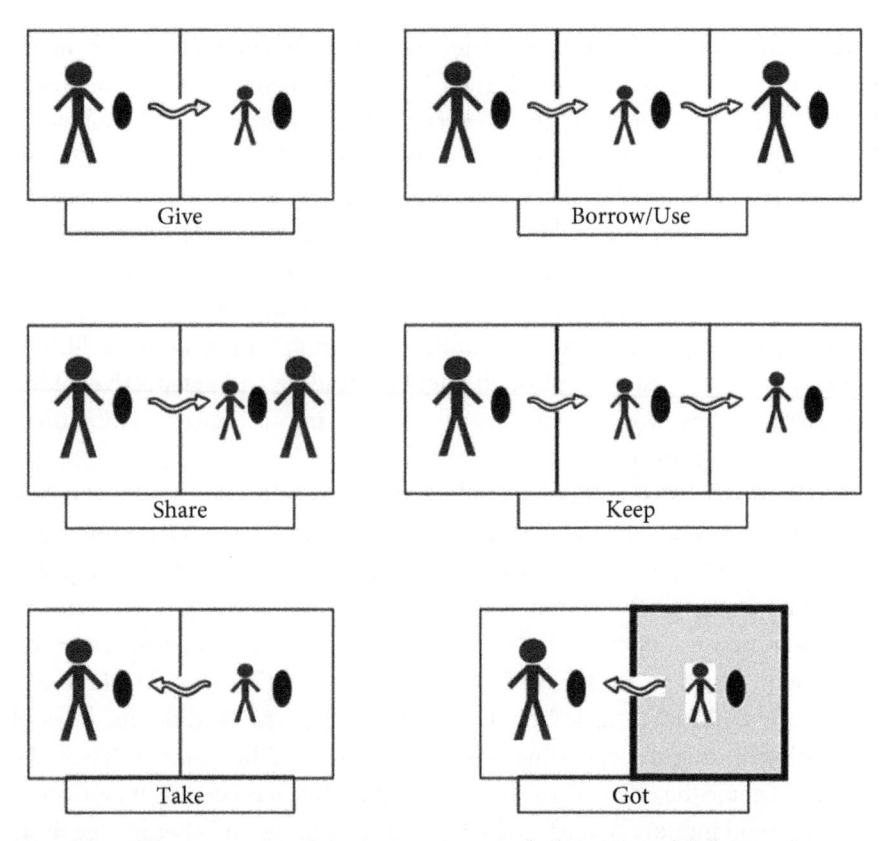

Figure 5.1. The perspectival representations underlying one child's use of six related verbs at around the second birthday. Each verb comprises two or three sequential "moments of attention" (one per rectangular box). The larger person represents the adult, the smaller person represents the child, and the oval represents an object. The arrow represents the causal force and direction of object movement. For example, *give* is the situation in which the adult causes an object to go from herself to the child. The shaded box indicates that the designated word profiles only the end state of the process.

Second are cases of perspectivizing more narrowly defined in which the child is construing the exact same situation from different attentional perspectives. For example, this same child, in the months following her second birthday, had several pairs of words representing polar opposite perspectives. For instance, she could talk about the lamp being "over" her head or her head being "under" the lamp. She could talk about someone "giving" an object to someone or that someone "taking" it. She could talk about the same object as either a *train* or a *toy*, or the same animal as a *dog* or a *pet*. She could talk about something she liked as being *good* or else being *better* than something else. She could talk about both *going home* and *coming home*, and she knew the contrast between *here* and *there* and *I* and *you*. And she knew that the same person on different occasions was a *woman* or a *teacher*. These are all instances of what I have been calling attentional perspective (not conceptual perspective, as characteristic of children after three years of age, as we shall see in the next chapter, which involves the child coordinating the different perspectives simultaneously and understanding how they relate to one another). But still, in both profiling and perspectivizing the child has at her disposal alternative ways of construing one and the same physical situation and chooses one of them strategically to fit the current communicative situation, including the listener's perspective.

The skills of joint intentionality involved in word learning and representation become especially clear when we compare what children are doing with what so-called linguistic apes are doing (Tomasello, 2017). With appropriate training from humans, great apes can learn many human words (in a gestural or graphic medium). But they do not have the same kind of perspectival representations as human children: no pronouns, no deictic terms like *here* versus *there*, and no choosing among alternative words to assist their listener's comprehension. It is thus the apes who are doing something more like mapping sounds onto things in an associative manner, that is, without the structuring of joint intentionality. They are in a human cultural world and exposed to its cultural tools, but they are mastering them with skills of individual intentionality only. In contrast, children are employing their skills of joint intentionality to acquire the conventional use of linguistic symbols for coordinating intentional states with others. As they are acquiring the use of their culture's linguistic symbols, children's ways of conceptualizing the world are thus being channeled in particular directions (as Vygotsky never tired of stressing). It should nevertheless be noted that toddlers are mastering these cultural tools without fully appreciating their conventional nature. An understanding of conventionality awaits collective intentionality at three years of age, when they will understand that only people from the same linguistic community participate in conventional agreements.

5.3.2. Grammatical Constructions as Role-Based Schemas

The process of acquiring the grammar of a language is similar in many ways to the process of acquiring words. But to appreciate the similarity one must appreciate the notion of a linguistic construction—which replaces the more familiar notion of a grammatical rule. A linguistic construction is an organizational pattern in the use of linguistic elements in a single utterance, a schema if you will. For example, if someone says "The modi is mibbing," we English speakers know that the modi is doing something; if someone says "The modi is getting mibbed by a blicket," we know that the modi is having something done to it by a blicket; and if someone says "The dax is mibbing the simba a blicket," we know that the dax is somehow transferring a blicket to the simba. We know the general meaning of these linguistic constructions even without knowing the meaning of any of the content words because the constructions themselves—that is to say, the pattern of how the words and other linguistic elements are organized into roles—are meaningful (Goldberg, 2003). The role-based schemas that constitute grammatical constructions have their origins in collaborative activities in which different participants play different roles.

The way young children acquire the grammar of their language, then, is similar—but also interestingly different—to the way they acquire words, that is, through the basic processes of intention reading and pattern finding. Thus, on a particular occasion the child must comprehend what the adult is saying to her in a multi-word utterance by reading his overall communicative intention, and then also, ideally, engage in some kind of "blame assignment" of how the different elements in that construction are contributing to the overall communicative intention (what role each is playing). For example, if an adult says to the child "The dog wants a bone," to comprehend that utterance as intended the child must understand both the overall communicative intention and the role that each of the elements is playing in it (e.g., the word "want" indicates the dog's desirous state, the phrase "the bone" indicates the desired object, etc.). Then, to become productive in using the construction with novel linguistic material, the child must find and master the pattern of use across other "similar" utterances. For example, perhaps the adult now says that the dog wants other objects or that other agents have other intentional states toward other objects, and the child discerns the organizational pattern. Linguistic constructions are thus abstract (or partially abstract) schemas for expressing communicative intentions using multiple words organized in particular ways. The most general point is that both word learning and construction learning involve processes of intention reading and pattern finding, albeit instantiated in different ways.

The linguistic constructions that toddlers first comprehend and use are those indicating the quasi-propositional situations with which they are already familiar from communicative pointing, namely: (i) the various motions and changes of state of inanimate objects (e.g., objects appearing, disappearing, reappearing, changing locations, changing states, etc.); and (ii) the various actions of intentional agents (e.g., people doing things, going places, manipulating objects, interacting with other people in various ways, etc.) (Slobin, 1985). Toddlers start out with very simple and concrete constructions organized around particular words, especially verbs. Learning verbs is key for learning grammar because the meaning of the verb *smile* includes a smiler; the meaning of the verb *kick* includes a kicker and a thing kicked; and the meaning of the verb *give* includes a giver, a recipient, and a thing given. Children's earliest grammatical constructions are thus what have been called verb island schemas or constructions (Tomasello, 1992), which have as their central element a particular verb that structures the roles involved. From the outset, these roles are generalized across particular participants to yield such verb island constructions as *Bring X; Y give X to Z; More X; Y gone; Z broke; X is kissing Y*, etc. Evidence for this organization is provided by both observational and experimental studies. For example, if one teaches a 23-month-old toddler a novel object word, *toma*, she can immediately use that word to fill the slots of many of the verbs she already knows; but if one teaches that same toddler a novel verb, *daxing*, she is unable to combine it with any other words productively at all—because she has not heard how this verb organizes its participant roles and she does not, at this point, have any verb-general abstract constructions (see Tomasello, 2003, for a review). The process of re-representation can easily abstract across the different objects that play a particular role in a construction, but abstracting across the organizational structure of the construction as a whole requires analogy, which is more complex.

From two to three years of age, toddlers gradually build up their grammatical competence. But they do not do this with any across-the-board abstract rules; rather, they do so initially in piecemeal fashion with each verb island construction separately and independently. For example, a 22-month-old might learn to say *Draw X* to indicate what she is drawing or plans to draw. Then gradually over the next year she learns to indicate the agent doing the drawing, the instrument with which she is drawing, and the medium on which she is drawing, saying in the end things like "I draw flowers on paper" (Lieven et al., 2003). Since she is building up each of her verb island constructions independently, the roles are verb-specific as well. That is to say, the 28-month-old knows that the one doing the drawing is said first, before the verb *draw*, and the thing being drawn is said after. But because the child has not generalized across verbs, it is not appropriate to talk about *subject* and *object* or even *agent* and *patient*; rather, the child is marking the "drawer" and the "thing drawn," and in a similar manner for each of her verb island constructions (Akhtar, 1999). These are thus roles tied to particular verbs, and each role is marked with what we may call a second-order symbol, such as word order or other syntactic marking. I call them "second-order" symbols because they do not symbolize a concrete referent but rather the role a concrete referent is playing in the schema/construction as a whole. It takes children some time to learn that such second-order symbols can be used productively with novel material since they indicate abstract roles in a complex schema.

The constructions in adult language go beyond these early-learned schemas in providing myriad ways of both profiling and perspectivizing what is occurring. Thus, an adult speaking to a child about an event may say either "The toy is spinning" or "Papa is spinning the toy," construing the event either from the perspective of the affected object (the toy) or the agent of the action (Papa). Moreover, the participants in events may be indicated in different ways grammatically depending on the common ground between adult and child, for example, the adult could refer to "the toy," "the toy we bought this morning," or "it," depending on how specific she needs to be in the context. Different languages do these things in very different ways, and so the child's task in moving beyond her very earliest role-based constructions is to find patterns in the way that the adults in her culture do them, learning new and more complex constructions in the process. Later in the preschool years, children will be able to say such things as "The toy was spun by Papa" (passive construction), or "It was Papa who spun the toy" (cleft construction), or "I think Mommy spun it" (propositional attitude construction) as needed in different communicative circumstances. So, just as in acquiring their early words, in acquiring their early grammatical constructions toddlers are construing the world in different ways via profiling and perspectivizing.

Again, the upshot is that children learn to use linguistic constructions in the ways that adults use them by reading their intentions in particular situations (in the context of joint attention), and then finding patterns in adult usage across contexts via processes of constructive thinking and re-representation. And again a brief comparison to "linguistic" apes is instructive (Tomasello, 2017). These apes can learn to string together their words (if that is what we should call them), sometimes with preferred orderings, and sometimes to create new meanings. But what they are totally missing is the pragmatic, or shared intentionality, dimension of grammatical competence. For example, "linguistic" apes do not:

- profile or perspectivize their reference to things in their common ground with the listener to help her identify the intended referent (e.g., no noun phrases with articles, no pronouns, and no relative clauses);

- use second-order symbols such as case markers or word order to mark for the listener grammatical roles indicating who is doing what to whom (i.e., their ordering preferences are not being used for communicative purposes);
- use any devices for indicating for listeners what is old versus new versus contrasting information (e.g., intonation or stress on new information); and
- choose constructions based on profiling or perspectivizing for particular contexts (e.g., they do not contrast "I broke the vase" with "The vase broke").

The key theoretical point is that human linguistic constructions are created with adaptations for the recipients' knowledge, expectations, and perspective in mind—this is the pragmatic dimension of grammar based on skills of shared intentionality—and it would seem to be precisely the uniquely human aspect.

Grammatical constructions provide young children with a symbolic format for conceptualizing role-based schemas instantiating the relational-thematic-narrative dimension of human experience. They thus depend on an understanding of role-based collaboration and the quasi-propositional structuring of the contents of communicative intentions. This representational format will become even more important for young children after three years of age as they learn propositional attitude constructions with mental state terms (e.g., "I believe/hope/doubt that it is raining outside") which symbolize in one schema both intentional states and their propositional contents. But even before three years of age, the way that toddlers think about the world is shaped in the relational-thematic-narrative direction both by their engagement in collaborative activities with others and by their comprehension of the basic grammatical constructions of their language.

5.3.3. Discourse as Joint Attention to (and Perspectives on) a Topic

Despite their impressive skills with words and linguistic constructions, toddlers are very poor conversationalists. Conversing with a partner on a shared topic requires children to transform their skills of joint attention to the perceived world into skills of joint attention to the linguistically represented world—that is, to jointly focus with a partner on a topic of conversation—and moreover to do so across conversational turns. Although 2-year-olds can reply to questions that adults ask or make comments on statements that adults make, it is very challenging for them to carry on a conversation in which they establish a joint topic with an adult and then each of them makes relevant comments about that topic in turn. And it is almost impossible for them to carry on an extended conversation with a peer.

But as they approach their third birthdays, toddlers become ever more capable of carrying on conversations, even with peers. Conversations are of crucial importance in cognitive development as they force children to coordinate and even negotiate perspectives with others. Thus, a child at a zoo might say to his mother, "Look at the deer," and the mother might reply "What big antlers," and then the child might make another comment. They are each expressing various perspectives on a common topic that they believe the other will find interesting or relevant. O'Madagain and Tomasello (2021) refer to this process as "joint attention to mental content" because in using linguistic symbols interlocutors express their perspective on a situation in a public medium that enables joint attention to that perspective with a partner, who might then give their perspective on that perspective. Discourse with others in a conventional language displays mental states for public scrutiny.

Joint attention to mental content plays a crucial role in children's coming to understand not just attentional perspectives but conceptual perspectives in which one and the same entity is perspectivized simultaneously in different (sometimes incompatible) ways. In the next chapter, we focus especially on situations in which children assert a propositional statement of fact—such as "That is a deer"—and others dispute that assertion by saying something like "You are wrong; it is a moose." In such cases, the child's understanding of a situation—in terms of her conceptual perspective of what is objectively the case—becomes an object of contention and negotiation. This will be critical for 3- to 4-year-old children coming to understand that they, as well as others, might be wrong about things (as we shall soon see).

5.3.4. Cognition for Linguistic Communication

Since Chomsky, it is commonplace to think of language as a kind of module, isolated from other cognitive processes and skills. This view is reinforced by the treatment of language acquisition in most developmental psychology textbooks, which introduce the process by abruptly abandoning the theoretical vocabulary of cognition, social cognition, and communication for the theoretical vocabulary of phonemes, morphemes, lexicon, and syntax. But whatever else it is, language is not a module. Learning to use a language in the way that other people use it requires all of human children's most powerful cognitive and social-cognitive capacities. And once a significant amount of language has been learned, it scaffolds and even enables many of children's other most impressive cognitive achievements.

The alternative I am proposing here is that toddlers' skills of linguistic communication develop in the context of their skills of collaborative activities in general and cooperative-referential communication in particular. Toddlers could not learn to use a natural language in adult-like ways without an understanding of the roles and perspectives inherent in joint intentional activities, as well as the ability to coordinate perspectives with others in gestural communication in attempts to draw the attention of others to quasi-propositional facts about the world. The essence of linguistic representations—as opposed to nonlinguistic cognitive representations—is that they construe or perspectivize a situation in a preestablished way. And the power of grammatical constructions lies in the ways they organize the construal of situations into (perspectivized) role-based schemas at various levels of abstraction. The fact that linguistic apes learn "words" without distinct perspectives and linguistic "constructions" without the marking of grammatical roles supports the view that what makes human linguistic communication unique in the animal kingdom is precisely its embeddedness in more encompassing processes of shared intentionality, with its species-unique conceptualizations of social roles and perspectives. Figure 5.2 is a highly oversimplified depiction of this embeddedness, proceeding from basic processes of joint agency at the bottom to processes of cooperative/referential communication (e.g., pointing) in the middle to grammatically organized communication at the top two levels.

A natural language may thus be seen as a structured inventory of cultural tools for coordinating with the intentional states of others for certain social purposes. This inventory has been built up over cultural-historical time in a society to serve the needs of a wide variety of individuals in a wide variety of communicative circumstances. In mastering their culture's language, children's conceptualizations and thinking about the world are thus led in particular directions. This is quite simply because to communicate effectively with adults, children must learn to think like adults, including in the use of adult communicative conventions. But children do not master a language by simply reproducing adult utterances; they find patterns in adult language use. They do this in the same basic way that they form generalizations and theories in other domains of their experience, that is, via processes of constructive thinking and re-representation that serve to coordinate and abstract across the different uses

96 AGENCY AND COGNITIVE DEVELOPMENT

```
                    ┌─────────────────────────────┐
                    │  GRAMMATICAL CONSTRUCTIONS  │
                    │   w/grammatically marked roles│
                    │     & discourse perspectives │
                    └─────────────────────────────┘
                         ↗                ↖
        ┌──────────────────────┐   ┌──────────────────────┐
        │  ROLE-BASED SCHEMAS  │   │ PERSPECTIVAL SYMBOLS │
        │   as conventionalized│   │   as conventionalized│
        │ linguistic constructions│ │ linguistic representations│
        └──────────────────────┘   └──────────────────────┘
                  ↑                          ↑
            ┌──────────────┐          ┌──────────────┐
            │    ROLES     │          │ PERSPECTIVES │
            │ in quasi-propositional│  │ on quasi-propositional│
            │ communicative situations│ │ communicative situations│
            └──────────────┘          └──────────────┘
                     ↖                ↗
                    ┌──────────────────┐
                    │   JOINT AGENCY   │
                    │ w/roles & perspectives│
                    └──────────────────┘
```

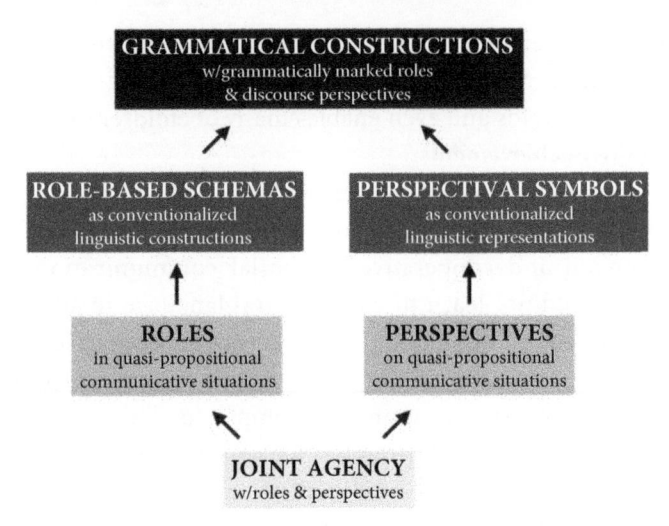

Figure 5.2. From bottom: basic capacities for joint agency with its roles and perspectives (light gray box). These are foundational for cooperative-referential communication with the pointing gesture with their use of roles and perspectives in communicative intentions (medium gray boxes). These are in turn foundational for children's earliest perspectival symbols and role-based schemas in a conventional language (darker boxes), as well as more complex constructions that mark grammatical roles and discourse perspectives with second-order conventions (black boxes).

of words and constructions that they observe in the social interaction and linguistic communication around them.

The mastery of linguistic symbols and constructions thus requires a complex set of cognitive processes and skills that, in turn, facilitate and shape children's other cognitive processes and skills. Most prominently, mastering a conventional language may impact children's cognitive development by: (i) making available to them a representational format that supports conceptualizations ever farther beyond the iconic representations of infants; (ii) providing them with conceptualizations and representations that others preceding them in the culture have found it useful to formulate; and (iii) connecting their knowledge across disparate cognitive domains. I will return to these proposals and, more generally, to the role of language acquisition in cognitive development, in Chapter 8 after we have considered development during the preschool years in the next two chapters. Language both depends on and is a major shaper of human cognition.

5.4. Toddlers' Joint Agency and Intentionality

At the same time that toddlers are becoming individual intentional agents who think, plan, and make either/or decisions in the medium of imaginative cognitive representations—as well as understand causality and intentionality and actively test hypotheses about how these work in particular cases—they are also becoming able to form with others joint agencies within which they collaborate and communicate in species-unique ways. In their collaborative activities, they form with a collaborative partner joint goals and intentions, and they pursue these in the context of joint attention and common ground. In their cooperative communication using the pointing gesture, they actively share and coordinate their actions and attention with a communicative partner, producing and comprehending recursively structured intentions and inferences in the process.

Operating in joint agencies engenders three new sets of cognitive processes. First, joint agency and intentionality lead toddlers to cognitively represent the world perspectivally. Beyond the imaginative representations of intentional agency in general, coordinating action and attention in joint intentional collaboration and communication empowers toddlers to conceptualize and represent proto-propositional situations from some imagined perspective (among other possible perspectives). Second, joint agency and intentionality lead toddlers to operate on these perspectives with recursive inferences. In cooperative/referential communication coordinating and communicating effectively requires them to consider the partner's perspective and how it relates to their own perspective (relating to their perspective, etc.). And third, joint agency and intentionality lead toddlers to an understanding of roles. This brings into existence for them the thematic-relational-narrative dimension of human experience in which entities are constituted by their role in some overarching activity or structure. As these novel forms of cognitive representation, operation, and conceptualization are developing, toddlers are using them to construct new types of perspectival concepts, knowledge, and theories—via the self-regulative processes of constructive thinking and re-representation—in many specific cognitive domains. This all sets the stage for the construction, beginning at 3–4 years of age in early childhood, of the kinds of multi-perspectival conceptualizations that underlie the objective and normative worlds of the cultural group (see Chapters 6 and 7).

In this period from nine months to three years of age, toddlers are not only developing these new ways of representing and understanding the world, but they are at the same time acquiring the culture's pre-existing forms of symbolic/linguistic representation, already perspectivized and structured in role-based schemas for communicating experience in conventional ways. In using these linguistic conventions in discourse with others, toddlers engage in joint attention not just to concrete objects and activities but to the mental contents of linguistic expressions, about which interlocutors coordinate and negotiate their individual perspectives recursively. The fact that skills of joint agency and intentionality emerge during toddlerhood in synchrony with skills of linguistic communication implies to some scholars that language is somehow responsible for perspectives, recursivity, propositional thinking, and other aspects of uniquely human cognition. My own view, however, is more or less the opposite: whereas acquiring a conventional language boosts toddlers' skills with all these uniquely human cognitive processes, it is the exercise of joint agency and intentionality more generally—in both phylogeny and ontogeny—that creates them in the first place (see Chapter 8).

In all the world's cultures, all these species-unique skills of collaboration and communication first emerge in the months around the first birthday. This suggests the possibility that they are all part of a maturational capacity for joint agency and intentionality, which nevertheless, I hypothesize, requires participation in collaborative and communicative activities to actualize. One question is why this emergence occurs in conjunction with the emergence of skills of individual intentional agency. If we assume that over generations in human evolution there was a gradual "migrating down" of skills of collaboration and communication to younger ages, one hypothesis is that this is the point at which the migration had to stop because this is the youngest age at which the necessary cognitive prerequisites are in place. In particular, forming joint agencies requires (i) intentional agents who understand others as intentional agents, (ii) imaginative representations to conceptualize different perspectives and roles, and (iii) an executive tier with a workspace within which to compare and coordinate roles and perspectives—all characteristics of intentional agency in general. In this context, it is instructive to think of the emotion-sharing characteristic of young infants in protoconversations as an earlier-emerging pre-intentional manifestation of social sharing.

Despite all these new and powerful cognitive capacities for joint agency and intentionality, young toddlers still are not very competent in directing and regulating joint agencies themselves. Their skills of joint intentional collaboration and communication are skillfully practiced only with adults, who do most of the directing, regulating, and decision making. Evolutionarily, it may be that joint agency develops in human ontogeny before it is needed for collaboration and communication with peers because, in the context of cooperative breeding and childcare, toddlers need to collaborate and communicate with the multifarious adults who are supporting them (Hrdy, 2006). And so, for more than two years after infancy, toddlers continue to operate with many impressive cognitive capacities and skills but mainly or only when they are interacting with adults.

III
EARLY CHILDHOOD

Every function in the child's cultural development appears twice: first, on the social level, and later, on the individual level; first, between people (interpsychological), and then inside the child (intrapsychological).
Lev Vygotsky

NB: As I will use it, the phrase "early childhood" refers to the developmental period from three to six years. There is no universally accepted way to refer to children in this age range, and so I often use the American-centric term "preschoolers" (although this may be misleading in societies in which formal schooling begins either earlier or later than six years).

6
Metacognitive Agency and Multi-Perspectival Representations

In traditional societies age three is the age of weaning, and so now young children (even in societies in which they actually wean earlier) are more often on their own to make and regulate their own decisions. This growing independence creates myriad new challenges, including the need to coordinate collaborative roles and perspectives not just with adults but also with co-equal peers. It is thus no accident that age three is the age at which young children begin learning and representing some especially complex types of multi-perspectival concepts that require them to coordinate multiple conceptual perspectives simultaneously.

The working hypothesis of this chapter is that what accounts for these new cognitive capacities, and others, is that children from three to six years of age are in the process of becoming metacognitive agents who reflectively regulate their executive-tier conceptualizing, thinking, and decision-making. What makes this rational/reflective mode of operation possible is a new organizational architecture that builds on top of toddlers' executive tier of regulation a second-order executive tier, a metacognitive tier, of regulation. The emergence of this metacognitive tier—which first evolved in ancient great apes to facilitate decision-making in the face of intense contest competition—transforms toddlers' fairly limited processes of executive re-representation into preschool youngsters' extremely powerful processes of metacognitive re-representation, enabling them to reflect on and coordinate key aspects of their cognitive operations, including those that occur in their shared agencies and perspective-shifting communication with others. This metacognitive re-representation results in whole new types of cognitive representations and concepts whose common characteristic is that they incorporate, in one fashion or another, multiple perspectives simultaneously. These new types of conceptualization enable preschoolers to construct much more complex and adult-like concepts, hypotheses, and theories about their physical and social worlds.

I will not look at all of these complex and adult-like concepts, hypotheses, and theories; that is something done spectacularly well by the modern theory of cognitive development in many different domains of knowledge. Instead, I will focus only on a more general description of the metacognitive tier of agentive organization and how it structures new processes of decision-making, learning, and re-representation, thus making possible new types of multi-perspectival concepts and theories. I analyze four example concepts to illustrate the process. The focus in this chapter is on how this all works in preschool youngsters' conceptualizing and theorizing about the physical world, and in Chapter 7 the focus is on how this all works in preschool youngsters' conceptualizing and theorizing about the social/normative world.

6.1. A Hierarchical Model of Executive Regulation

As noted already in Chapter 4, contemporary research on executive function is quite diverse with much confusion in terminology. The main reason for this confusion is that different investigators have different goals, many of them focused on identifying individual differences in executive function in relation to outcomes such as school achievement and emotional adjustment. Much the same may be said about contemporary research in metacognition, which spans many diverse topics for many diverse research goals, including most prominently the goal of enhancing children's learning in various educational settings. Here I am concerned only with the most basic processes of metacognition and how they work.

In the current model, metacognitive processes just are executive processes; both function within a control system whose primary goal is to regulate behavioral decision-making. In developmental psychology, the term metacognition often singles out the fact that the child can "think about thinking" without referring at all to regulation. But the child is thinking about thinking in order to solve a problem, to learn a new skill, or to achieve a challenging goal, all of which require them to make a behavioral decision. Roebers (2017, p. 33) argues that what are called in the literature executive function and metacognition actually play quite similar roles in children's behavior and cognition: "Both are higher-order cognitive processes enabling an individual to operate flexibly and adapt efficiently to new and challenging tasks . . . [Both] similarly encompass dynamic and regulatory functions, which are utilized to optimize information processing of more elementary, first-order tasks." She notes that in the way they are studied in the current literature the two functions comprise different "subprocesses." Paraphrasing slightly to emphasize aspects relevant to the current account, for executive function these are such things as attention shifting, behavioral updating, and behavioral inhibition, and for metacognition they are cognitive monitoring and cognitive control (see also Carlson, 2023).

Executive function and metacognition are thus both processes of regulating—that is, monitoring and controlling—psychological functioning; they just differ in what they regulate and how they regulate it. In the current model, executive regulation employs cognitive processes such as thinking and planning to monitor and control action and attention, whereas metacognitive regulation employs second-order cognitive strategies to monitor and control the executive-tier cognitive processes themselves. (My paraphrasing of Roebers above was only to add in the words "attention," "behavioral," and "cognitive" in appropriate places to emphasize this distinction.) Note that I am not saying that everything researchers have referred to as executive function is about the regulation of attention and action, and everything researchers have referred to as metacognition is about the regulation of executive processes. I am proposing a different way of referring to things. To make clear that different way, I briefly compare the current model with Diamond's (2013) typology of the three major categories of executive function: (i) inhibition; (ii) working memory; and (iii) cognitive flexibility.

With regard to inhibition, two-year-old toddlers engage in proactive processes of inhibitory control, especially the inhibition of unchosen behavioral options. Although they are notoriously poor at inhibiting prepotent actions—for example, inhibiting eating the marshmallow in front of them for a greater reward later—toddlers nevertheless have at least some inhibitory control over at least some prepotent actions. They can also inhibit to some degree unwanted attention to distractions. Inhibiting unwanted actions and attention may be called **executive inhibitory control**, and children become much better at it across the preschool years (see Diamond, 2013, for a review). But beyond toddlers, preschoolers are also capable of monitoring and inhibiting processes of executive decision-making itself—what may be called **metacognitive inhibitory control**. For example, in some problem-solving contexts preschoolers question and inhibit an established belief/decision if new evidence suggests that it is not accurate or is likely to fail. It is also important in these control processes to distinguish reactive control (after executing an action) from proactive control (in anticipation of an action). I examine such processes further in the following section on metacognitive decision-making.

What is classically called working memory is best described as a kind of executive workspace where all kinds of cognitive processes (not just memory) take place. In two-year-old toddlers, this workspace may be called the **executive workspace.** It hosts cognitive simulations and the proactive planning of actions along with other forms of executive monitoring and control. For example, if a toddler were attempting to cross a stream by stepping on stones, she might survey the available stones and cognitively simulate a simple behavioral plan before setting out. The executive workspace expands greatly during preschool; thus, Völter and Call (2014) found that in a maze planning task four-year-old children were able to plan one step ahead, whereas five-year-old children were able to plan two steps ahead. But also during the preschool years a second executive workspace—the **metacognitive workspace**—emerges, and it hosts metacognitive strategies for regulating and coordinating executive-tier thinking and planning. For example, a preschooler might formulate a plan for stepping on stones to cross a stream, but then stop and metacognitively reassess that plan if new information comes to light (e.g., her brother falls into the stream while crossing). Then, she might compare the merits of that plan with those of a different cognitively simulated plan, such as using the bridge downstream. I examine in more detail the ability to evaluate and choose among plans—and also beliefs—again in the following section on metacognitive decision-making.

Cognitive flexibility is classically measured by tasks such as the Dimensional Change Card Sort task (Zelazo, 2006), and the basic finding is that two-year-old toddlers are capable of sorting objects into piles of similar objects (e.g., by shape or by color), but if they are then asked to sort the same objects in a different way (i.e., on the other dimension), they are inflexible and fail. Preschool children become ever more flexible and skillful in this task across early childhood, with initial competence starting at around three or four years of age. This task may be viewed as a challenge to toddlers' perspective-taking. Recall that toddlers are able to executively monitor and compare **attentional perspectives**: *what* different individuals are attending to. But with the ability to understand cognitive processes from a metacognitive tier, preschool children after three or four years of age come to understand **conceptual perspectives**: *how* individuals are differently conceptualizing one and the same thing simultaneously, which derives from young children's shared intentionality interactions with others in the context of a conventional language (Tomasello, 2018a). So toddlers are understanding (from an executive tier) single-focus attention whereas preschoolers are also understanding (from a metacognitive tier) potentially multi-perspectival conceptualizations. Cognitive flexibility as it is most often studied in the literature, then, requires preschoolers' metacognitive processes.

METACOGNITIVE AGENCY 107

The overall conclusion is thus that each of Diamond's three classes of executive function—inhibition, working memory, and cognitive flexibility—has an executive and a metacognitive version, depending on whether the target of monitoring and control is attention and action versus cognitive processes. But we may reconceptualize this typology as a single hierarchical system of agentive decision-making and action. Figure 6.1 illustrates the broad outlines of such a model, specifying: (1) *what* is regulated (attention and action in toddlerhood vs. executive cognitive processes in early childhood); (2) *how* it is regulated (e.g., via processes of action planning and executive inhibitory control in toddlerhood vs. metacognitive coordination and inhibitory control in

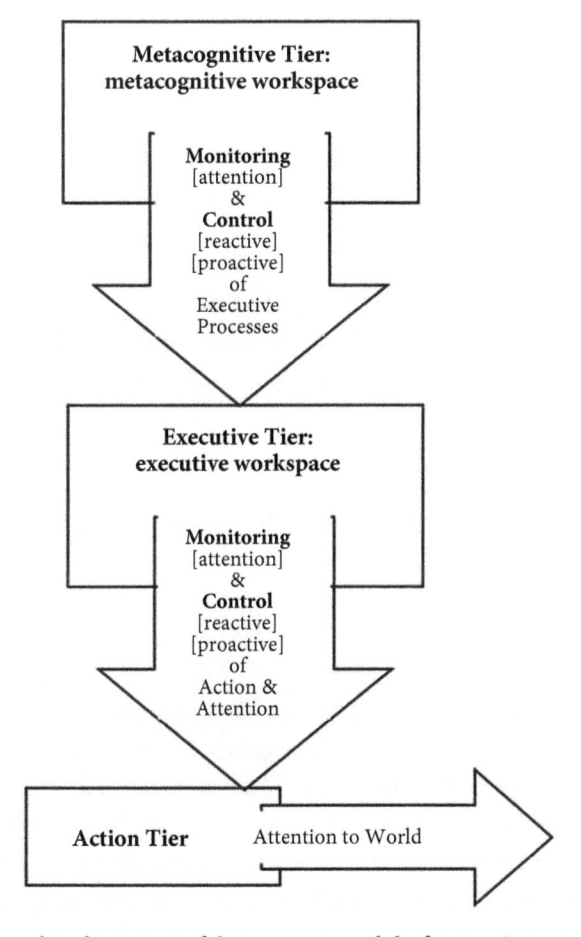

Figure 6.1. Graphic depiction of the current model of executive processes, including what is regulated, how it is regulated, and where, in what workspace, it is regulated (see Tomasello, in press).

early childhood); and (3) *where* it is regulated (executive workspace in toddlerhood vs. metacognitive workspace in early childhood). Although I know of no existing models of executive or metacognitive regulation that take precisely this form, there are hierarchical models of executive processes—albeit focused on different phenomena than the current model—in both the adult (e.g., Koechlin & Summerfield, 2007) and developmental (e.g., Zelazo, 2004, 2015) literatures. The specific mechanisms of executive and metacognitive regulation currently posited in the literature (often defined by cognitive task) reflect a focus on one or another application or sub-process in this overall regulatory architecture.

6.2. Metacognitive Decision-Making

Evolutionarily speaking, the "proper function" of both executive and metacognitive processes is the facilitation of behavioral decision-making. As seen in Chapter 4, toddlers and other mammals are able to executively monitor their actions and attention from an executive tier of regulation and then proactively think, plan, and make either/or decisions. But preschoolers now become able to metacognitively monitor these executive processes of thinking, planning, and decision-making and so reflectively decide among different possible either/or decisions, including revising already-made decisions and/or beliefs in the light of new evidence or new reasons.

6.2.1. Reflective Decision-Making

How animals and children use metacognition to make decisions is often studied using tasks of uncertainty monitoring. For example, when presented with a difficult discrimination or memory problem, many animal species and preschool children opt out and go for a safer alternative: they know that they do not know. But there is controversy over whether this actually requires metacognition in the strict sense of the term (e.g., see papers in Beran et al., 2012). The key issue in the current context is whether children younger than three years of age are able to metacognitively reflect on the decision-making process.

In two important studies researchers have posited metacognitive decision-making in two-year-old toddlers. In one study described in Chapter 4, when two-year-olds were uncertain about their ability to solve a behavioral problem, they recruited a parent for help (Goupil et al., 2016). In a similar study, also described in Chapter 4, when two-year-olds did not see where an adult hid a toy—so they were uncertain where it was—they actively looked around a barrier to gain needed information (Call & Carpenter, 2001). These two studies are sometimes characterized as involving metacognition under the interpretation that the toddlers "know that they do not know" something. However, a different interpretation is that the toddlers in these studies are not metacognitively monitoring what they do and do not know, but rather they are executively monitoring what they can and cannot do: whether proceeding with a planned action is or is not likely to be successful in reaching the goal. In the view of Goupil and Proust (2023), monitoring behavioral uncertainty in this way is not monitoring a thought but rather monitoring a *feeling*. That is, the toddlers are executively monitoring a feeling of uncertainty as they try to choose an action, not metacognitively monitoring the executive-tier cognitive processes that they are using to make that decision. Goupil and Proust actually refer to this type of uncertainty monitoring as a procedural form of metacognition, that is to say, a form that focuses not on cognition proper but on ground-level processes of action and attention. I would thus characterize these two studies—and others like them in which children monitor behavioral uncertainty—as concerned with the executive supervision and control of action and attention.

Then, beginning sometime after three years of age, with the development of the metacognitive tier of agentive functioning, young children become able to metacognitively monitor and control not just the feeling of behavioral uncertainty but the cognitive processes involved in executive decision-making itself. This takes place in two different forms. One takes place within the agent's mind, as it were, as young children plan and evaluate possible executive-tier decisions before making a final decision, or perhaps reassess things after a decision has been made if new information becomes available (skills shared with other great apes). The other takes place between agents' minds, as it were, as young children coordinate decisions with others in joint or collective agencies (skills not found in other apes since they are not capable of creating joint or collective agencies).

Within minds, O'Madagain et al. (2022) gave both great apes and human children (three and five years of age) the opportunity to visually locate the best food in a situation at location X. The subjects did this, indicating their belief/decision by choosing that location (though not receiving the food as a result). Then, they were exposed to new information that called their initial belief into question: some new information suggested that the best food might be in location Y. Subjects then had the possibility to seek further information (or not) that could either confirm or disconfirm their initial belief. Many apes then actively sought more information to resolve the discrepancy between their original belief and the new information, by looking again into location X (and perhaps Y) to check their initial judgment so as to make the best decision. The apes were in this case metacognitively assessing their executive decision *after they had already made it* (which distinguishes the demands of this task from those of the two toddler studies reviewed above); they were reflecting on the belief guiding their decision in the light of newly obtained information and discerning the need to possibly revise that belief and so decision. If this is indeed what they were doing, it is important because attempting to causally diagnose problematic decisions before they are behaviorally executed fulfills a standard criterion for reflective/rational agency, and it clearly is metacognitive.

Like the apes, the human children in this task questioned their own belief and actively attempted to double-check it—but only at five years of age. The children at three years of age just went with one or the other choice without double-checking. However, in a second study, O'Madagain et al. (2022) provided subjects with discrepant information in a different manner. Specifically, the subject made an initial choice—again without actually receiving anything as a result—and then a conspecific entered and indicated a different choice. In this case, the apes did not double-check their initial choice, presumably because they did not compare the perspectives of themselves and the peer. In contrast, in this situation the human children actively double-checked their initial choice, and they did so even at three years of age! This suggests that, in contrast with apes, young children find different perspectives emanating from social partners to be more salient indicators of the need for belief revision than new information emanating from the physical world. In their individual decision-making, young children are especially attuned to discrepant social perspectives, which prompts them (i.e., more strongly than physical evidence) to metacognitively reflect on and revise their beliefs and so decisions.

6.2.2. Joint Decision-Making with Peers

Between minds, preschool children for the first time begin to mentally coordinate with co-equal peers to make truly joint decisions in joint agencies. Whereas two-year-old toddlers can coordinate their ongoing actions and attention with others (mostly adults), preschool children can plan and coordinate their actual *decisions* with others, including peers. The process of coordinating not just actions but decisions is studied formally in game theory in what are called coordination games. A well-known coordination game—and one that may have special significance in human evolution—is the stag hunt (Tomasello 2014, 2016). In the classic parable, I am hunting alone for hares when I spy a stag, which is more and better food but which requires collaboration for capture. You are in the same situation, and so it behooves us both to drop our pursuit of hares and collaborate to capture the stag. The problem is that neither of us can be certain that the other will choose to go for the stag (maybe our partner did not see the stag). Chimpanzees do not perceive the stag hunt as a dilemma: they just go for the stag and hope the other will follow. But four-year-old children perceive the dilemma and so before leaving the hare they hesitate and/or communicate with their partner to make a joint decision (Duguid et al., 2014).

Four- and five-year-old preschoolers can even coordinate their decisions in situations in which the possibility of communication is eliminated, that is, in games of so-called pure coordination (which great apes cannot do; Duguid et al., 2020). That is, they are able to coordinate their decisions on a common choice if there is some salient feature of one of the choices—e.g., one is red while all the others are white—which they can metacognitively predict will be a salient decision for their partner, whom they know is attempting to metacognitively predict their decision as well (Grüneisen & Tomasello, 2015a). Moreover, children in this same age range are even able to plan a coordinated decision in a joint problem-solving situation by each partner predicting which tool each of them must choose in her role—and then coordinating their respective choices before acting (Warneken et al., 2014).

In more discourse-based studies of decision-making with peers, pairs of three- and five-year-olds are able to coordinate a joint decision by metacognitively comparing their different beliefs—and even reasons for their beliefs—through perspective-taking discourse and joint reasoning. For example, in one study peer partners had different information from different sources about what food some novel creatures typically ate. To resolve the issue, they metacognitively discussed the validity of the evidential sources from which they each obtained their information (hearsay vs. direct observation) and came to a reasoned joint decision as a result (Köymen & Tomasello, 2018). In these joint problem-solving situations peers coordinate not just their actions but their decisions, which requires each of them to metacognitively monitor both their own and the partner's beliefs, as well as their respective reasons for their beliefs (see Köymen & Tomasello, 2020, for a review of studies of joint decision-making and reason-giving in preschool peers).

Preschool youngsters are able to coordinate their decisions with others, in the current hypothesis, because they are now operating with a new metacognitive tier of functioning that enables them to conceptualize and socially coordinate executive-tier cognitive processes such as beliefs and reasons, with which they, from three to four years of age, are operating. The Vygotskian hypothesis is that it is precisely this kind of social coordination of beliefs, reasons, and decisions with others that is the original source of preschool children's individual cognitive flexibility and conceptual perspective taking, as they reflect on and internalize the social process into an internal dialogue to deliberate on their own about what to do or believe. The O'Madagain et al. study described earlier (in which children reflectively examined their own beliefs at a younger age in the face of a discrepant social perspective than discrepant physical information) is generally consistent with this view. Also supportive is the study of Köymen et al. (2020) in which adults trained three-year-olds in a kind of "meta-talk" discourse about reasons, evidence, and their validity, and this led the children later to engage in more skillful joint decision-making with peers.

6.3. Metacognitive Learning and Re-representation

Metacognitive monitoring and control lead to some new processes of learning and cognitive re-representation. Indeed, much of the research on metacognition in both preschool- and school-age children is aimed at identifying metacognitive strategies that can help them learn more efficiently in educational settings.

6.3.1. Metacognitive Learning: Belief Revision

Learning by preschool children acting as metacognitive agents takes place via many of the same processes used by infants and toddlers; that is, in many cases they learn something new and just overwrite whatever, if anything, was already there. But they also at this age begin to engage in more systematic and strategic metacognitive learning, a.k.a., belief revision, in which existing knowledge is represented as beliefs that are based on evidence and always subject to revision. For example, when preschoolers have strong evidence for a belief about some causal relationship (there is a strong "prior"), seeing a single discrepant event does not lead them to revise their belief, whereas seeing that same discrepant event when existing evidence for a prior belief is weak often does lead them to revise their belief (Kimura & Gopnik, 2019). Further, preschool children only revise their beliefs when an interlocutor gives good rather than bad reasons, supported by sufficient evidence, and they even take into account meta-reasons (reasons for reasons; Schleihauf et al., 2022). Overall, during the preschool years, children are becoming better able to compare the evidence for their existing beliefs with various kinds of newly obtained evidence metacognitively and strategically and so to become ever better Bayesian learners. In addition, as noted above, in joint decision-making with peers, conflicts of belief and subsequent reason-giving lead to reason-based belief revision.

During early childhood, youngsters also become much more selective and strategic in seeking just the evidence they need in testing hypotheses. Although even toddlers can formulate and test hypotheses in a general way, preschoolers become able to metacognitively assess just what information they are missing and formulate metacognitive strategies for obtaining it (see Butler, 2020, for a review). As a special case, preschool children are also becoming much more skilled and strategic in interrogating adults to get just the information they need (see Ruggeri, 2022, for a review), again suggesting an ability to metacognitively assess just what information or evidence is needed. Further along these lines are the classic studies of metamemory in which preschoolers develop strategies—either with or without adult assistance—for helping themselves to recall things in the future that they anticipate they would otherwise forget (e.g., Schneider, 1999; for a more recent study see also Bulley et al., 2020). And preschoolers are also beginning to be able to form metacognitive strategies for learning to learn and for thinking reflectively in both social and individual contexts (skills most often studied in school-age children in middle childhood; Schraw et al., 2006). Although it is not clear whether social interaction, including pedagogy, is necessary for the development of these metacognitive learning skills, it is clearly facilitative (Kuhn & Dean, 2004).

6.3.2. Metacognitive Re-representation

The most important type of learning in modern cognitive-developmental theory is Bayesian learning, which goes beyond classical forms of learning in requiring additional cognitive processes. Most obvious is the abductive creation of hypotheses to be tested. The problem, as both Carey (2009) and Xu (2019) point out, is that formal models of Bayesian learning operate with a fixed set of hypotheses. But assuming there is not a lifetime supply present at birth, this raises the question of where these hypotheses come from. Carey's (2009) Quinian bootstrapping is one attempt to address this question, and Xu's (2019) constructive thinking is another. In the current model, Bayesian hypotheses are generated most often as the child engages in constructive thinking to determine either what to do or how something works in the face of one or another form of uncertainty. Evidence in one direction or another may lead the child to re-represent her knowledge.

For Bayesian learning to work, then, the child needs processes of constructive thinking and re-representation. These processes reorganize conceptual material in two basic ways that often occur together: (i) the child can abstract/analogize commonalities across cognitive content (vertical integration), and (ii) the child can coordinate and/or synthesize disparate pieces of cognitive content into one coherent conceptualization (horizontal coordination). As an example of vertical integration, a child might learn to call her family pet a *dog*, and then hear it called an *animal*, which seems discrepant. Observation of further uses helps to specify the relation between these words. Then as she observes multiple uses of similar word pairs such as *woman* and *person*, she might, through an act of constructive thinking, abductively hypothesize and re-represent both of these as instances of the more encompassing notion of superordinate/subordinate relation (via what Gentner, 2010, calls analogy or structure mapping). Upon encountering *car* and *vehicle*, even without multiple learning encounters, she might now use her new understanding as a hypothesis to test on this new pair of words. The other, horizontal dimension of constructive thinking and re-representation enables children to coordinate conceptualizations by synthesizing them into composite concepts. For example, to construct the concept of *natural number*, the child must integrate or synthesize her existing concepts of cardinal number and ordinal number, which might occur as she observes people counting a set of objects sometimes using cardinal and sometimes using ordinal designations. Or if the child is puzzled by her friend searching for an object in one place when she herself can see it in another, she might resolve the discrepancy by coordinating the different perspectives involved—hers, the friend's, and an objective perspective—and so make a general distinction between the subjective *beliefs* of individuals and the objective situation independent of individuals.

Constructive thinking and re-representation operate not only in the context of learning, as in these examples, but also via "offline" reorganizations of the child's existing knowledge base (i.e., outside of relevant decision-making and action). These reorganizations occur as the child reflects on discrepancies and inefficiencies in her knowledge base that need resolution. Just as a scientist might reorder her thinking while riding a bicycle in the countryside, a child might see some new relation between conceptualizations outside of any directly relevant context (while doing other things, so to speak). This reorganization process is related to what has been called in agent-based modeling "computational rationality," in which the individual assesses the efficiency of her behavioral decision-making from one or another executive tier (e.g., Lieder & Griffiths, 2020). Re-representation thus sometimes operates as a kind of offline computational rationality as the individual reflects on and reorders her existing knowledge base in order to make it a more efficient and effective source of information for future decision-making and learning.

All of these re-organizing processes require an executive workspace, and so only operate after infancy. After infancy, they operate in the same basic manner throughout development, but the outcome is very different because the conceptual resources with which and on which they operate change over age. Consequently, we must distinguish between the **executive re-representation** of toddlers and the **metacognitive re-representation** of preschoolers based on the kinds of conceptual material they are re-representing and the ways they are re-representing it. Whereas toddlers' executive re-representation operates on attention-based perspectival concepts, three- to five-year-old youngsters' metacognitive re-representation operates, in addition, on material involving conceptual perspectives. The metacognitive preschooler is thus able to coordinate different conceptual perspectives in ways that makes their relationship clearer, perhaps synthesizing them into some new concepts. For example, whereas toddlers can understand different attentional perspectives as alternative ways of construing the same objects on different occasions (e.g., as dogs or as animals), preschool children after three years of age can coordinate multiple ways of categorizing the same objects *simultaneously* based on different conceptual perspectives—and so synthesize an understanding of hierarchies of classification (e.g., the class of dogs as hierarchically included in the class of animals).

Of special importance in this process is children's re-representation of their social coordination, perspective-taking, and self-regulation in shared agencies, as these enable the construction of uniquely powerful types of perspectival and objective/normative concepts. As described already in Chapters 4 and 5, one- and two-year-old toddlers' constructive thinking and executive re-representation operate not only on action and attention (e.g., in our spooning dirt example), but also on joint action and joint attention and their associated roles and perspectives. This leads to toddlers' construction of attention-based perspectival concepts, sequentially applied, especially in their acquisition of the culture's linguistic symbols and constructions. In addition, preschool youngsters after three years of age can metacognitively reflect on their discourse interactions with others in which differing conceptual perspectives must be coordinated *simultaneously*. Such metacognitive reflection on shared intentional interactions and conceptual perspectives enables preschool children to construct both concepts such as *natural number* that simultaneously coordinate cardinal and ordinal conceptual perspectives (as we shall see in the coming section), as well as concepts such as *belief* and *reason* that simultaneously coordinate the differing conceptual perspectives of multiple individuals by synthesizing and re-representing them in a single concept (as we shall see in the coming chapter).

My contention is that whereas much cognitive development during early childhood may be attributed to either maturation or learning, in addition much of it may be attributed to developmental processes of constructive thinking and metacognitive re-representation, as the child reflects on and re-organizes her executive processes of thinking, planning, and decision-making from a metacognitive tier of functioning. Of special importance to preschool children's cognitive flexibility is reflection on their shared intentional interactions with others in which they negotiate and exchange differing attentional and conceptual perspectives, which may then be internalized, in Vygotskian fashion, into individual perspectival construction. This species-unique dimension of metacognitive functioning leads to a whole host of new cognitive skills, representational formats, and concepts.

In the coming section I look at four exemplars of the new ways that preschoolers conceptualize the world multi-perspectivally, all from the physical/mathematical world. In the chapter that follows I deal with multi-perspectival concepts from the social/normative world.

6.4. Some Multi-Perspectival Representations

One of the foundational tenets in modern cognitive science is that concepts—even those indicated by a single linguistic item—can only be understood in the context of a wider body of knowledge (or "theory," in those approaches that stress a scientific metaphor). The concept of *knuckle* can only be understood in the context of knowledge about fingers and hands (in the context of animate bodies); the concept of *mother* can only be understood in the context of knowledge about reproduction (and possibly families); the concept of *electron* can only be understood in the context of knowledge about atoms (and matter). A crucial dimension of concepts is their role or place in a larger body of knowledge.

In this section I examine four concepts (in their respective knowledge domains) that classically have been considered important in young children's cognitive development. Each of these concepts merits a chapter of its own to do it full justice, but here I have two more circumscribed aims. First, I want to show that in all of these domains a relatively major transformation occurs in preschool children's understanding of the target concepts at three to four years of age. And second, I want to show that in each case a major cause of the transformation is children's emerging ability to coordinate, and in some cases synthesize, multiple conceptual perspectives on the same cognitive content simultaneously (and so to re-represent it). I will not attempt a full developmental explanation for any of these conceptual achievements but will simply seek some explanatory perspective on each of them by considering why they seem to be out of reach both for great apes—who function metacognitively but do not operate with perspectival representations—and for toddlers—who operate with perspectival representations but do not function metacognitively.

6.4.1. Object Classes

Human toddlers, and a number of other primate species, categorize objects based on perceptual similarities, as evidenced by their selectively touching similar objects sequentially or even placing similar objects together into spatially segregated groups (e.g., Langer & Killen, 1998; see Mandler, 2007, for a review). Human toddlers, and some specially trained great apes, also categorize objects in communication, as evidenced by their use of category terms in symbolic communication (Savage-Rumbaugh et al., 1998). Categorizing objects into groups is thus a very natural cognitive process for both apes and human toddlers. But Inhelder and Piaget (1964) and many subsequent researchers have found that toddlers are basically incapable of classifying a single set of objects in multiple different ways simultaneously. Indeed, early Piagetian research found that even preschool-age children struggle with multiple classification. However, by stripping away extraneous task demands contemporary research has found important skills of multiple classification in preschool children—both successive and simultaneous—identifying a key transition at around three to four years of age.

The most common task to measure young children's skills at successive multiple classification is, again, the Dimensional Change Card Sort task (DCCS;

Zelazo, 2006). As noted earlier, in this task children are required first to sort cards on one dimension (e.g., color) and then immediately sort them by another (e.g., shape). Early research tended to show that three-year-old children had trouble classifying objects in a second way (see Doebel & Zelazo, 2015, for a review). But subsequent research employing more child-friendly versions of the task have found that performance is quite good at 3–3.5 years of age, whereas it is very poor at 2.5 years of age (e.g., Blakey et al., 2016). So, age three would seem to be the key age of transition for successive multiple classifications. The most common task used to assess simultaneous multiple classification is matrix completion. In this task, children must find the missing object in a matrix created by crossing two dimensions, for example, placing a red triangle in the missing space defined by the convergence of a red vertical dimension and a triangle horizontal dimension. Again, early studies showed that three- and four-year-old children struggle with this task, but Podjarny et al. (2017) designed a more child-friendly version and found that both three- and four-year-olds were quite competent. Interestingly and importantly, Podjarny et al. (2022) administered child-friendly versions of both a successive and simultaneous task of multiple classification and found that young children were consistently better at the successive version.

What explains this relatively sudden competence at three to four years? Based on a series of nine studies using the DCCS (as well as a review of relevant literature), Zelazo et al. (2003) concluded that children's performance was not best explained by developments either in memory or in inhibitory control. The best explanation was what they called a "redescription account" (championed most prominently by Perner & Lang, 2002), which attributes growing success to young children's developing ability to appreciate multiple perspectives on the same object(s) at the same time. But why does this new ability emerge only at around three or four years of age? In the current account, the obvious reason is that this is the age at which the new metacognitive tier of regulation emerges, and this enables children to re-represent all of the simple categorization activities in which they have been participating for several years already. So perhaps on one occasion the child labeled an object a "bird" and then on another occasion noted that it was a "cardinal," or on one occasion she singled out the ovals from a group of blocks and on another occasion singled out the blue ones. These acts create discrepancies in that the same object is conceptualized as different things on different occasions. Constructive thinking and re-representation on the metacognitive tier use these kinds of discrepant experiences as the raw material to coordinate and perhaps synthesize contrasting conceptual perspectives on the same entities to enable multiple classifications of the same object in different ways for all kinds of creative purposes, first successively and then, in certain contexts, simultaneously. Relatedly, it is well-known that toddlers fail to understand that the same person may be both a teacher and a mother whereas preschoolers understand that the same object can be different conceptual things (Markman, 1989).

This account does not specify the many kinds of cognitive skills and learning experiences that go into the process of multiple classification. It only establishes that (i) multiple classification involves the coordination of conceptual perspectives, and (ii) multiple classification skills emerge developmentally at around the same age as the metacognitive tier of regulation at three to four years of age. In this view, great apes are not skillful at multiple classification—which they are not, as far as we know—because they do not understand conceptual perspectives in the first place, since these emanate from intercourse with others in shared intentionality interactions.

6.4.2. Physical Symbols

Toddlers not only manipulate and categorize objects, they also use them in acts of pretense. Although many such acts have been called "symbolic," most theorists agree that in their early pretense toddlers are not seeing the object in two ways at the same time; the pretense act is "quarantined" from the toddler's representation of the object as a real object (Leslie, 1987).

However, in some cases children face a problem of "dual representation." In a series of classic studies, DeLoach (summarized in 2004) found that toddlers have a difficult time conceptualizing an object simultaneously as a real object and as a symbol. In the basic task, children are shown a dollhouse and given training in how it is set up to represent a real room next door: the same furniture in the same places, and so forth. An adult then hides a small toy dog in a particular place in the dollhouse and tells the child that she is going to hide a large toy dog in that same place in the real room next door. Across many studies, DeLoache found dramatic differences in the performance of two- and three-year-old children in this task, with only three-year-olds seeming to comprehend what was going on. DeLoach's explanation relies on what she calls "dual representation." To succeed in this task, children have to see the dollhouse not only as a physical object with its own set of affordances, but also as a symbolic representation of the other room (which two-year-olds struggle to do). In support of this interpretation, two-year-old children succeed in the task when they are led to believe that the adult is capable of magically transforming the small room into the large one. In this case, they understand themselves to be searching in the same room on both occasions, just before and after a size transformation—with no issue of dual representation—so they have no problem locating the large toy dog.

One might worry that mapping onto one another the many items in a furnished room with those in a dollhouse simply overtaxes toddlers in terms of complexity. But other studies that rely only on single objects as representations of other objects have found that two-year-olds still struggle in a way that three-year-olds do not. For example, Tomasello et al. (1999) set up a game in which an adult requested that the child push one of several objects down a slide to her. When the adult requested the object by name, or by a habitually associated gesture (e.g., hammering for a hammer), both two- and three-year-olds were quite good. But after having established the game in this way, if the adult requested the object by holding up a small replica object, only children approaching their third birthdays were skillful. In terms of production in this study, again only the three-year-olds were good at using a conventional artifact as a symbol for another conventional artifact (e.g., a book as a hat). It is perhaps worth noting that sounds or gestures may be used as symbolic representations in language even by young toddlers because they do not create a problem of dual representation as do physical objects.

Once again, then, my proposal is that children can use physical objects as symbols only when they can conceptualize the same object as different things simultaneously, and they can do this only after three years of age, that is, only after the metacognitive tier of regulation has emerged.

6.4.3. Natural Number

Many animal species have the ability to compare the relative quantities of items and choose the larger one. This ability has presumably evolved in the context of efficient foraging for food, especially under the pressure of competition with conspecifics. In addition, humans have evolved the ability to construct the concept of *natural number*: the series of numbers that young children typically use to count. Although some great apes have been trained in some competencies with natural numbers, they only master some aspects of the concept—as indicated by the fact that after they have been trained to understand Arabic numerals up to 7 or 8, when they are now trained on the Arabic numeral 9 it takes them just as long to learn it as it did the initial numerals. They do not seem to get the idea of a recursively structured set (see Call & Tomasello, in press, for a review).

As for humans, even six-month-old infants can discriminate different quantities, even quite large ones if they are sufficiently different (Xu & Spelke, 2000; see Call, 2000, for the same finding in great apes), and they can even keep track of exact quantities if only one or two objects is involved (Wynn, 1992; see Mendes et al., 2008, for the same finding with great apes). Toddlers and apes also do two other things relevant to the concept of natural number: they place numerals in order (ordinality) and they identify the numerosity of very small sets (e.g., indicating something like "two shoes" and "two apples") (cardinality). In Piaget's (1952) original analysis, to understand the concept of natural numbers, children have to coordinate the ordinal perspective and the cardinal perspective: they have to understand, for example, that the number 4 is simultaneously larger than the number 3 but smaller than the number 5 (ordinality) and that all sets of items in 1-to-1 correspondence to the fingers of one's hand have the same numerosity(cardinality).

Gelman and Gallistel (1978) provided a detailed account of young children's development of the number concept and the order in which they acquired the various component concepts. The endpoint was what they called the cardinality principle, namely, the understanding that when counting objects (respecting 1-to-1 correspondence) the numerosity of a set of objects corresponds to the last numeral said in the series. Arguably, mastery of the cardinality principle in counting requires a coordination of ordinality (which already involves seeing the same number in a series as both larger than its predecessor and smaller than its successor) and cardinality. Wynn (1990, 1992) devised a simple test of young children's understanding of the cardinality principle: she simply asked young children to give her, for example, five pennies out of a larger set. They typically counted, but the toddlers below three years of age were not able to coordinate their counting procedure with the actual objects in front of them in accordance with the cardinality principle: they handed over random numbers of pennies. Starting at about 3.5 years of age, however, young children were able to master this so-called Give-N Task flexibly with many different numbers and many different types of objects. In a detailed analysis of children's mastery of the different counting principles, Cheung et al. (2022) confirmed 3.5 years of age as the typical age of mastery of the cardinality principle in the Give-N Task. Their conclusion is that this is the age at which children come to coordinate their mastery of the number series, 1-to-1 correspondence, and cardinality judgments. They claim that "learning to coordinate all three principles represents an additional step beyond learning them individually" (abstract).

Again, then, the claim is that among the many factors that account for children's development of a concept of natural number, a critical factor is the development of a metacognitive tier of functioning in which they can coordinate different conceptual perspectives—in this case, something like an ordinality perspective and a cardinality perspective—via constructive thinking and metacognitive re-representation. Cross-cultural research (e.g., Frank et al., 2008) suggests that the emergence of a metacognitive executive tier of functioning and the ability to take multiple perspectives on the same item are not sufficient conditions for understanding the natural number concept; there also must exist counting practices in the culture of a sufficiently robust nature. But, again, my only claim is that metacognitive coordination of different perspectives on the same material is a necessary part of the process, and it only emerges after three years of age.

6.4.4. Causal Nets

Toddlers understand quite a bit about causal events, both when their own actions cause external events and when external events cause one another (see Chapter 4). And they even express this in their language, as many two-year-olds use the word "because" to talk about the relation between all kinds of events, for example, saying that a car can't go "because it doesn't have any gas" (Bloom & Capatides, 1987). Toddlers also demonstrate a basic understanding of causal events in novel experiments, including how they can intervene to influence them (Gopnik et al., 2001; Sobel and Kirkham, 2006; see Chapter 4).

But beyond these straightforward causal inferences, Gopnik and Wellman (2012) provide a theoretical analysis of children's more complex types of causal reasoning in terms of chains of Bayesian inferences using directed graphs. This framework helps to formalize the way that toddlers and young children make many different kinds of causal inferences involving such things as screening off irrelevant information, making retrospective inferences, predicting the effects of a behavioral intervention, and so forth. As part of this analysis, Gopnik and Wellman examine the question of how young children create hierarchical Bayes nets (including so-called overhypotheses) concerning more abstract patterns of causal structure that apply across whole classes of similar events. For example, from learning causal facts about the physiology of squirrels, rats, and dogs, young children may construct more general causal principles about such things as the mammalian circulatory system. Of special importance in the current context, thinking in terms of directed graphs requires children to understand that the exact same event can simultaneously be both an effect of some previous event and a cause of some subsequent event, as, for example, the movements of a tool are simultaneously both the effect of an agent's actions and the cause of effects in the world.

Experimentally, and again using the blicket machine paradigm, Schulz et al. (2007) showed young children a number of specific causal events (e.g., in the pattern X activates Y, and Y activates Z, but X does not activate Z directly). In this case, Y is a mediating event that is both an effect and a cause. They found that four-year-old children could infer more general rules of this type based on fairly small amounts of evidence. (To my knowledge, younger children have not been tested in this paradigm.) Also relevant in this context are studies of young children's abilities to engage in counterfactual causal reasoning. For example, using a blicket machine, Nyhout and Ganea (2019) demonstrated for young children both overdetermined trials (two blocks they knew to be causal placed on the machine) and single-cause trials (one block they knew to be causal and one block they knew to be non-causal placed on the machine) and asked them what would have happened if one of the two blocks had not been placed on the machine. Four-year-olds' performance was above chance on both trial types, whereas three-year-olds' was not. Arguably, counterfactual causal reasoning requires the child to simultaneously consider a particular event as both a possible cause and a possible non-cause in the same scenario.

One more time, then, the argument is simply that the nature of children's causal understanding changes in significant ways at around age three or four, with an important advance being the child's emerging ability to represent a single event simultaneously as both a cause and an effect in the overall sequence. One important factor contributing to this developmental advance is the emergence of a metacognitive tier of psychological functioning on which the child can reflect on her more local and concrete causal thinking to construct more abstract and coordinated paradigms of causal reasoning via metacognitive re-representation.

6.4.5. Summary

My very brief accounts of these four domains of physical and mathematical knowledge are not intended to be anything close to complete. They are intended to support only two conclusions: first, that in all of these domains important developmental progress occurs at around three or four years of age, and second, that examination of the nature of what occurs at this age suggests

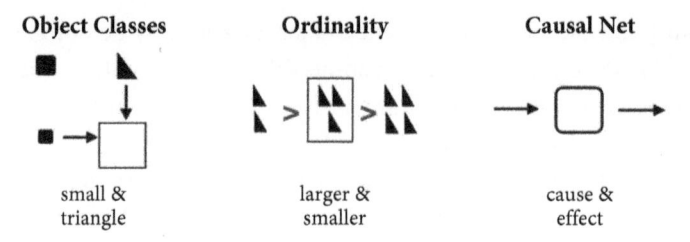

Figure 6.2. Graphic depictions of three multi-perspectival concepts from early childhood. Arrows depict different perspectives; the box indicates a concept requiring coordination of two perspectives.

an important role for metacognitive processes. More specifically, my hypothesis is that the observed developmental progress at three to four years of age in all these domains is due, in large part, to a process of metacognitive re-representation in which the child reflects on her already existing knowledge and concepts and then abstracts, coordinates, and synthesizes them in various ways metacognitively depending on the specifics of the task and domain. A key version of this process in all the examples considered here is synthesizing different conceptual perspectives on the same material to create multi-perspectival representations. Figure 6.2 graphically depicts three such concepts.

This process presupposes large amounts of learning in each specific case. If preschoolers were simply reflecting on knowledge from toddlerhood, there would be little developmental progress. Metacognitive re-representation works in concert with a continual process of learning new things. The claim is simply that the emergence of the metacognitive tier plays a necessary role in this progress as it opens up for children new formats of cognitive representation based on new capacities for the coordination of perspectives. The falsifiable prediction is quite specific: one could engage in all kinds of training and pedagogy with one- and two-year-olds, and that would not suffice in any of the current cases for them to master the knowledge and representations that come more or less naturally after the emergence of the metacognitive tier at three to four years of age.

6.5. Metacognitive Organization and Concepts

Metacognitive organization first emerged in early great apes to facilitate executive processes of thinking, planning, and decision-making in the face of intense contest competition. Most importantly, it enabled individuals to reflect on already-made decisions and improve them—and also to eliminate inconsistencies and inefficiencies in their knowledge base via offline processes of metacognitive re-representation—before acting. This form of metacognitive organization with re-representation first emerges in young children at around three years of age, but from the beginning it operates differently than it does in other apes because the conceptual materials on which children are operating metacognitively includes attentional and conceptual perspectives as they are manifest in both linguistic representations and collaborative interactions involving mental coordination with others.

From about three to four years of age, then, young children are operating with a three-tiered system of agentive organization, two of which involve executive regulation. This conceptualization of two tiers of executive regulation—executive and metacognitive—is not in conflict with existing models; it simply emphasizes different aspects and conceptualizes them in a somewhat different way. Adopting the definitions I have proposed here leads to the not-so-controversial conclusions that: infants before nine months of age mostly lack executive processes (with the exception of some kind of global inhibition); toddlers after nine months of age begin to executively regulate their actions and attention (although at first not so skillfully); and preschool youngsters after three years of age begin to metacognitively regulate their cognitive processes of thinking, planning, and executive decision-making (although, again, at first not so skillfully). What is perhaps more controversial is the claim that the emergence of these executive and metacognitive processes reflects naturally occurring cognitive-developmental reorganizations that transform not only children's behavioral decision-making but also their capacities for constructing new types of knowledge and concepts, especially those that coordinate and synthesize multiple perspectives on a single event or entity simultaneously.

The claim is that this new format of multi-perspectival cognitive representation enables young children to begin moving beyond the relatively immature theories and concepts of toddlers toward the more mature theories and concepts of actual adult scientists. Thus, for the concepts we have examined here:

- **Object Classes** enable the hierarchical classification of natural phenomena—with different classifications for different theoretical purposes—that are a bedrock of scientific theories. Hierarchies of classes are needed for everything from classifying geological materials to identifying biological genera and species.

- **Physical Symbols** enable the kinds of graphic representations that scientists use in their everyday work, from electrical and engineering blueprints to the kinds of graphical depictions of data that populate virtually every scientific publication.
- **Natural Number**, as Galileo so famously put it, constitutes the "language of nature" for all of the natural and most of the human sciences.
- And since a major aim of all the empirical sciences is to identify causal relations among objects and events in the world, phenomena with any degree of complexity require explanations in terms of **Causal Nets** in which each node of a network is both effect and cause, for example, in biological ecosystems.

Of course, preschool youngsters' versions of these basic scientific tools and concepts have a long way to go before they are fully adult-like. My only claim is that preschoolers' new multi-perspectival format of cognitive representation constitutes a crucially important component in the cognitive architecture that they will eventually need if they are to begin doing real science. Their new multi-perspectival format of cognitive representation also creates the possibility of concepts and theories with an objective and/or normative dimension, as we shall now see.

7
Collective Agency and Objective/Normative Representations

As preschool youngsters are becoming ever more competent and independent as individual agents, they are at the same time becoming ever more dependent on the social group in which they live. To find their way in the group children must not only make many of their own decisions, but they must also conform to the way things are conventionally and normatively done in the group—such that they are able to collaborate and communicate effectively with all its members, including those they have never met before. Preschool youngsters are gradually coming to identify with a collective "we," encompassing not just collaborative partners of the moment but everyone who identifies with "our" ways of doing things.

This new way of being reflects a suite of evolved human capacities for participation in the collective agency and intentionality of a cultural group. Its ontogenetic roots lie in toddlers' earlier capacities of joint intentional collaboration and communication with individual roles and perspectives. But then, at three to four years of age, preschool youngsters scale up this more local way of interacting to the entire cultural group and its conventions, norms, and institutions. Again there is a dual-level structure, in this case comprising the collective agent "we" of the cultural group in contrast to the individual with her particular role and perspective. The common ground role ideals manifest in the earlier joint agencies now become normative cultural ideals toward which "we" in the collectivity strive: what "we" all agree anyone *should* do. The individual perspectives manifest in earlier joint attention now become contrasted with a kind of universalized "we" perspective: what "we" all agree anyone *should* believe (i.e., the "objective" situation). This scaling up from joint agency/intentionality to collective agency/intentionality thus relies on universalizing inferences to collective ideals that transcend the particular preferences and perspectives of individuals, leading committed group members to think in terms of what it is *necessary* to believe and do.

When this recognition of collective ideals combines with the metacognitive, multi-perspectival way of thinking, the result is two especially important sets of multi-perspectival concepts. Epistemically, children after three or four years contrast the subjective perspectives of individuals with the (idealized) objective situation, as a key part of, for example, the concept of *belief*. Morally, they contrast the subjective preferences of individuals with the ideally right thing to do, as a key part of, for example, the concept of *fairness*. Both objective and normative concepts, then, arise from the tension between the perspectives and values of the individual and the collective ideals of the "we" with which she identifies.

7.1. Collective Agency and Intentionality

Evolutionarily, collective agency and intentionality refer to the behavioral and cognitive competencies that evolved in modern human individuals to facilitate the pursuit of collective goals via collective decisions within the collective common ground of the cultural group. This group-mindedness does not just mean associating with the denumerable individuals of the group, but rather identifying with an idealized cultural collective understood as anyone (past, present, or future) who would be one of us, as demonstrated in shared behavior, appearance, and values. Cultural group-mindedness is about identifying with the "we" of the collective.

There are some precursors to group-mindedness both in infancy (viz., the expectation that individuals who look alike and/or congregate together will behave in similar ways; Powell & Spelke, 2013) and in toddlerhood (viz., a tendency to imitate people who are similar to the self; Buttelmann, et al., 2013). But after three or four years of age, youngsters begin to understand that they themselves are members of a larger cultural group. Thus, it is only at this age that children assigned to a "minimal group"—constituted only by a group name (e.g., "the green group") and/or characteristic clothing (e.g., green T-shirts)—begin to show in-group favoritism and other forms of parochialism (see Dunham, 2018, for a review). Importantly, it is at this same age that preschool youngsters begin not just to imitate others when it is useful to do so, as do other apes, but actually to conform to their group even when it is against their own best individual judgment, which other great apes do not do (e.g., Haun & Tomasello, 2014). And when faced with a coordination problem in a novel situation with in-group peers, youngsters' natural tendency is simply to conform to what others are doing (Grüneisen et al., 2015b).

From a cognitive point of view, crucial in all of this is the concept of collective or cultural common ground (a.k.a., the mutual knowledge of the group). Eighteen-month-old toddlers already create with a partner in a joint agency personal common ground based on things that the partners have personally experienced together in joint attention. As elaborated in Chapter 5, this personal common ground is critical for the forms of collaboration and cooperative communication in which toddlers participate. But collective (or cultural) common ground is based not on shared experiences among individuals but rather on an understanding of one's participation in an idealized cultural group, all of whose members have presumably had many of the same experiences because they have grown up in the midst of the culture's shared behavioral practices and conventions—and, critically, *everyone knows that everyone knows this.* Like personal common ground, this new understanding is thus recursively structured ("I know that they know that I know . . . "), but unlike personal common ground, it applies universally to anyone and everyone—past, present, or future—who would potentially be one of "us."

Important in the current context, this understanding of what is and is not in cultural common ground emerges only at around age three. For example, many two-year-olds are relatively proficient in their native language, but only three-year-olds understand that the linguistic symbols they are using are shared by everyone in the linguistic community and not by others (Diesendruck, 2005). Similarly, three-year-olds (but not two-year-olds) base their interactions with an in-group stranger on the assumption that she is familiar with important objects of their presumably shared culture (e.g., Santa Claus in the U.S.) but not novel toys (Liebal et al., 2013). It is during early childhood, then, that children begin to understand cultural practices and knowledge as the collective common ground of a collective "we."

Group-mindedness and collective common ground are crucial to uniquely human psychology in many ways. But for children's cognitive development during early childhood, of special importance are the ways that these structure a kind of universalization or idealization of perspectives and roles (initially developed in joint intentional interactions with other individuals in toddlerhood). Preschoolers do not just attempt to align and coordinate with their partner's attentional perspective on the situation but rather with the objective situation as "we" collectively and universally know it. And preschoolers do not just understand that the individual roles in a shared agency have mutually known standards of ideal performance but rather standards that represent the objectively right thing to do as "we" collectively and universally know it. Participation in the collective agency of a cultural group, then, involves new forms of convention-based conformity based on cultural common ground and idealized normative standards, which provide the raw material for preschoolers to begin constructing multi-perspectival knowledge and concepts that incorporate either a universalized epistemic ideal (of objective knowledge) or a universalized normative ideal (of moral rightness or wrongness).

7.2. "Objective" Knowledge and Concepts

Nonhuman primates know many things about the world, but they do not distinguish the subjective perspectives of individuals from an objective perspective independent of individuals. Two-year-old toddlers take a first step down this path. They jointly attend with a partner to "the same" object, event, or situation, but also realize that the two of them may have different attentional perspectives, leading to the insight that my partner's perspective is just like mine, is equivalent to mine, and so must be equally respected. Then, sometime after three or four years of age, preschoolers construct from this attentional perspective taking the notion of conceptual perspectives in which one and the same entity is construed differently—as an apple or a fruit, for example—depending on how one chooses to conceptualize it. As they begin to appreciate the universalizing ideals of collective intentionality, children then metacognitively construct and re-represent the notion of a universalized objective perspective independent of individuals, a notion that contains within it the possibility of an individual having an erroneous perspective (or belief).

Reinforcing this new way of looking at things is adult pedagogy. From adults in the culture, children are constantly exposed to the objective perspective on things as expressed in linguistically asserted generic propositions such as "Snakes are dangerous" (Gelman & Bloom, 2007). In pedagogy, adults intend that children trust them and their authority—as representatives of the culture—to know what objectively is the case and accept this knowledge, which children have a natural tendency to do (Csibra & Gergely, 2009). Although infants already recognize when an adult is communicatively addressing them (which that adult sometimes understands as pedagogy), children only begin to understand and behaviorally react to pedagogy as expressing culturally generalizable and universal knowledge at around three years of age (Butler & Tomasello, 2016).

Skills of metacognition (exercised on a metacognitive tier of functioning in a metacognitive workspace), and collective intentionality thus lead to a new understanding of knowledge itself. The notion of conceptual perspective implies that an agent may choose to conceptualize a single situation in different ways for different purposes. But this sometimes creates for preschoolers what Perner et al. (2003) call "perspective problems," which are problems because preschoolers (but not toddlers) understand that the world is objectively a certain way, which can bring different conceptual perspectives into conflict (though this is often only apparent and can be resolved). Here I review several of these perspective problems with the modest goal of establishing that (i) perspective problems only emerge after three to four years of age, and (ii) the key to their solution is the metacognitive coordination of different perspectives, one of which is an idealized objective perspective.

7.2.1. Conceptual Perspective-Taking

The classic task of visual perspective-taking is Piaget's famous three-mountains task, but it turns out that this is an especially difficult task for many reasons. Almost as famous is the turtle task, in which the child and adult view a sketch of a turtle from opposite sides of a table, and the child is asked, in various ways, how the turtle appears to each of them (i.e., right-side-up or upside-down). This task forces the child to compare how she sees the turtle to how the adult sees it at the very same time, which, for children after three years of age, creates a perspective problem that they only solve after four years of age (Flavell et al., 1981).

A pair of recent studies have helped to identify what makes this seemingly simple task so difficult. Moll and Meltzoff (2011) gave three-year-olds experience with a yellow color filter which changed the apparent color of things behind it. After this experience, children were seated at a table across from an adult and presented with two identical blue objects. One was placed out in the open and the other was placed on the child's side of the yellow color filter, so that it appeared green to the adult. Looking straight ahead, the adult then requested either "the blue one" or "the green one." Even though the objects appeared identical to the children, they chose the correct one from the adult's perspective in both cases.

In this study children could simply look at the two objects and determine which one of them appeared green to the adult, and, importantly, they could do this without contrasting perspectives with one another. Moll et al. (2013) therefore modified the task so that children had to identify the different colors of *one and the same object* as it appeared from their own perspective and from the perspective of an adult at more or less the same time (they were asked "How does it appear to you? And to me?"). Now the three-year-olds were pulled to the real—saying the object appeared blue both to the adult and to them—and it was only four-year-olds who understood that the exact same object that was blue for them was green for the adult across the table. Although there is nothing inconsistent about something appearing blue to me and green to you if color filters are involved, it is likely that three-year-olds' problem is that they think of the color of an object as an objective attribute: something cannot be simultaneously green (all over) and also blue (all over). And so, given that three-year-olds are good at imagining what others see in straightforward situations, it is likely that their emerging sense of an objective perspective—what color the thing really is—and/or their difficulty in coordinating two conflicting perspectives simultaneously, interfere with their ability to take the conceptual perspective of the other when they must explicitly compare it to their own.

Great apes should not be able to understand conceptual perspectives in situations such as this one, and indeed they do not. Karg et al. (2016) had two chimpanzees compete over two food sticks. While the subject could see that both were the same size, to the competitor one appeared larger than the other (because disproportionate parts of them stuck up above a barrier for the competitor). Across two studies of this type (also including preschool children), the conclusion was that "chimpanzees solved the task by attributing their own preference to the other, while children truly understood the other's mistaken perspective" (abstract). Although great apes reflect on their experience metacognitively, they have no notion of perspective—either attentional or conceptual—because they do not understand themselves and their partner to be in joint attention to an objective situation.

It is thus striking that two-year-old toddlers and apes can imagine what others are experiencing in many situations—and toddlers can even take their attentional perspective—but neither is capable of understanding in adult-like ways situations in which different conceptual perspectives must be coordinated and compared to the objective situation. This understanding awaits skills of metacognition and collective intentionality at three to four years of age. It is therefore only after this age that children understand and operate with concepts such as *appears* or *seems*, which presuppose an objective situation with which a perspective contrasts.

7.2.2. Appearance-Reality

The classic appearance-reality task assesses children's understanding that an object which appears to be one thing is really another, for example, an object that looks like a rock is really a sponge (Flavell et al., 1981). Children struggle with this task also until around four years of age.

Moll and Tomasello (2012) modified the classic task to again distinguish cases with and without conflicting perspectives. In a first study they presented three-year-olds with a nondeceptive and a deceptive object: for example, a bar of chocolate along with an eraser that looked like a bar of chocolate. They then asked the children to point to the "real" bar of chocolate or else to "the one that only looks like" a bar of chocolate. Three-year-olds were mostly successful in identifying each object correctly. However, in a second study children of this same age were presented with *a single ambiguous object* and asked to point to which of two exemplars—an eraser or a bar of chocolate—this single object "only looks like" and what it "really is." The same children who were previously successful with two different objects were not able to correctly answer this pair of questions about a single object. Like the studies of visual perspective-taking, in the first task children only had to conceptualize an object in one way at a time (e.g., either as chocolate or eraser), whereas in the second task they had to conceptualize it in two different ways *simultaneously,* and these ways seemed to conflict conceptually (what the object *is*). Like the studies of visual perspective-taking, it is likely that the three-year-olds had trouble because they were invoking an objective perspective such that the object cannot be two things at the same time. Four-year-olds resolved the conflict—or else did not even see a conflict—through a new understanding of the situation that accommodated the different perspectives involved: objects can appear as one thing but actually be another.

Two studies have asked if great apes can distinguish appearance and reality. Karg et al. (2014) used a task that two-year-old children in the same study also passed, whereas Krachun et al. (2009) used a more difficult task that only four-year-old children passed. The problem was that in both cases—because apes cannot communicate in language—the procedure was that the ape first saw the true situation (e.g., larger food on left) and then things were changed so that they now appeared differently (e.g., larger food on the right). Some apes were able to keep track of the true situation despite the changed appearance, but this could be because they were simply able to keep track of the objective situation throughout the transformation. Though this is an interesting ability, it does not show that apes can simultaneously consider the same entity under two conceptual perspectives at the same time (which children do when they express that "It looks like a rock but it's really a sponge").

Again it is striking that toddlers and great apes can imagine what others are experiencing in many situations but neither seems capable of understanding situations in which different perspectives must be coordinated and compared to the objective situation. Again, this comes only with skills of metacognition and collective intentionality at around three to four years of age. Understanding situations of this type underlies concepts such as *appears* or *seems* or *only looks like*, which presuppose an objective situation with which some perspective contrasts.

7.2.3. Linguistic Aspectuality

Children younger than four years of age also have trouble reconciling the fact that the same object may be called *horse* or *pony* (Doherty & Perner, 1998) or, in a similar case, *horse* or *animal* (the classic class inclusion task). Again, in point of fact there is no conflict once one learns how linguistic labels work; one may call something an animal or a horse or a pony or a filly or a nag or a nuisance, all depending on how one wants to perspectivize the entity or situation for one's communicative partner. In linguistic philosophy, it is said that the same object is being construed under different descriptions or aspects.

But young children do not initially understand the situation in this way; they assume that an object's label is an inherent property such that there is only a single objectively correct label at one time, which creates the conflict (Markman, 1989). Recently, Rakoczy et al. (2015) tested young children for an understanding of aspectuality in a situation where there was potential for error. They first saw a trick object in one state and assumed it was a toy carrot and then later saw it in another state and assumed it was a toy rabbit, sometimes later being shown that they were different states of the same object. It was only after four years of age that children understood that an object they knew under two different descriptions—carrot and rabbit—could really be one and the same object. So, again, we may posit that the younger children thought that once something was under one description that defined what it was objectively, and it could not simultaneously be something else, whereas the older children understood that it was just a matter of two different conceptual perspectives based on how one thought about things.

To my knowledge, there is no evidence that great apes who have been trained in human-like systems of communication understand that one and the same entity may be conceptualized simultaneously under two different linguistic descriptions. Children's metacognitive (metalinguistic) understanding of linguistic aspectuality does not create specific new concepts, but rather leads them to an understanding of linguistic communication and how it functions to convey different attentional and conceptual perspectives on things, depending on one's communicative intentions.

7.2.4. Beliefs

Classic tests of false belief also present perspective problems, although in this case there is no solution in which both perspectives are retained as valid. In classic false belief tasks, the child converses with an adult about a situation they have witnessed together. A third party witnesses this situation also but then is absent when it changes. The adult then asks the child how the third party will behave in the new situation. Again, in this case three-year-old children do not appreciate the conflicting perspectives involved—they just assume that everyone knows the objective situation—whereas four-year-old children typically do (see Wellman et al., 2001, for a review). That is, the four-year-old child understands that the third party believes one thing, whereas the objective situation—as best she can determine it—is something else. Thus, to perform competently in this task—to judge that the third party has a false belief about the objective situation—the child must coordinate three perspectives: the third party's, her own, and an "objective" perspective about what is actually the case independent of what either of them thinks. (Note that it is common to equate the child's perspective with the objective situation because the child has seen the object moved; but for a mature understanding the child must recognize that, in principle, she herself could be wrong about the objective situation as well.)

There are several theoretical proposals about exactly what an understanding of false belief entails. Once again, my only concern here is to establish that whatever is the best explanation, it will have to include the child coordinating different perspectives with respect to one another and with respect to the objective situation as she understands it. In the current theoretical account, this is only possible from age three onward. (Note that several more child-friendly false belief paradigms find competence soon after the third birthday, e.g., Rubio-Fernandez & Geurts, 2013, and Carpenter et al., 2002.) Importantly, the so-called implicit measures of false belief understanding—at which toddlers and even great apes are successful (e.g., Onishi & Baillargeon, 2005; Krupenye et al., 2016)—do not require a coordination of perspectives. Although some theorists interpret these studies in terms of the subject's understanding that a third party has a false belief, in fact toddlers and apes only have to imagine and/or predict what that third party has experienced and will do in the situation. There is no need to relate that to what they themselves believe, much less to the objective situation as they understand it. They do not have to understand the agent as having a *false* anything; they just need to take her perspective (Tomasello, 2018a; see also Southgate, 2020, who notes that infants and toddlers are "altercentric" in that they do not attend to their own perspective at all in such situations).

And so the conclusion is that, while two-year-old toddlers (and apes) understand quite a bit about how others perceive and know the world, they do not have a metacognitive understanding of beliefs as distinctive mental states that may or may not correspond to the objective situation. They lack a metacognitive understanding of beliefs as distinctive mental states because they do not yet have the notion of objectivity deriving from collective intentionality, which is what generates the possibility of an erroneous belief. Understanding beliefs in adult-like ways leads to adult-like mastery of a whole host of "objective" concepts, that is, concepts that presuppose a comparison to some objective standard. These include mental state concepts such as *think, know,* and *believe,* as well as concepts such as *true, false, correct,* and *incorrect* (which children first master at around 3 to 3.5 years of age; Bartsch & Wellman, 1995), and also a number of pragmatic discourse markers such as "actually," "in fact," "supposedly," etc.

7.2.5. Reasons for Beliefs

In situations of joint decision-making, beliefs often need to be justified. Thus, if the experimenter in a false belief task were to challenge the child's judgment that the third party is mistaken about the object's location, the child would have to say something like "But he didn't see it moved" to explain the false belief. Indeed, philosophers often define knowledge as "*justified* true belief" because someone might have a belief that corresponds to the objective situation—the cookie is in the box—but this judgment is based on some crazy reason—the cookie vaporized from one location and seeped through the cracks into the box. Objectively accurate knowledge of a situation thus includes some understanding of the *reason* why the situation is the way it is, which serves to connect the current belief logically or causally with knowledge that is already established as objectively valid in the common ground that is shared with some larger "we." Reasons provided in the context of joint decision-making are intended to be dispositive because they are "objective" in this way.

Young children typically give reasons for why they believe what they believe only when queried or challenged by someone, prototypically in cooperative decision-making when participants must come to some common understanding of a situation. The earliest age at which young children provide coherent justifications for their beliefs is, once more, three to four years. These justifications are most often statements of fact that the reason-giver believes are objectively true and also believes that her interlocutor will believe as true because they connect either logically or causally to something they both believe in common ground to be true. Thus, if a pair of preschool children are trying to decide where to place a toy polar bear in a toy zoo, they might just say, "There is some ice," which is a sufficient reason for a joint decision because of the partners' shared belief that polar bears live on ice. This shared belief has very likely not been discussed between the two of them previously but is based on the cultural common ground that they both assume they share through common education (Köymen et al., 2014). Although language is the natural medium for expressing reasons, it is not strictly necessary. When three-year-olds are trying to decide together where to look for a toy animal that leaves tracks when it travels, they can justify their suggestion for where to look by pointing to the animal's tracks and where they lead. The adequacy of this justification depends on the common ground belief (established earlier in the experiment) that the animal leaves tracks wherever it goes (Köymen et al., 2020). The process is thus again one in which the child justifies her belief by connecting it (in this case causally: the tracks were caused by the animal) with common ground beliefs she shares with her partner.

Understanding and producing reasons for beliefs thus requires the young child to coordinate her own beliefs, her partner's beliefs, and the objectively correct beliefs that the two of them share in common ground. Once again, as in the other examples, this requires a metacognitive tier of functioning (with a metacognitive workspace and metacognitive re-representation), along with skills of collective intentionality. And once again this tier and these skills are operative only after three years of age. Children are now able to operate appropriately and skillfully with the concept of *reason* and with lexical items such as "why" and "because."

7.2.6. Developmental Explanations

I have proposed that for all this knowledge and these concepts one factor determining the age at which they appear is the emergence of the metacognitive tier of functioning at three years of age and another is the emergence of skills of collective intentionality that accompany this transition. But there are other factors involved, and because of this many of these concepts are only mastered at age four. Because it has been experimentally investigated so intensively, the best test case for assessing the specifics of such a multi-factorial explanation is false belief. It has been found that two important factors contributing to the age at which a particular child acquires the notion of false belief are certain aspects of language and executive function.

First, many studies have found correlations in children's various skills with language and their false belief understanding (see Milligan et al., 2007, for a review). Indeed, children growing up deaf and with limited experience in a conventional sign language are significantly delayed in their understanding of false beliefs (Woolfe et al., 2002), with deaf children who grow up with no experience in a conventional sign language failing at nonlinguistic false belief tasks even as adults (Pyers & Senghas, 2009). Experience with a language would thus seem to be necessary for coming to understand false beliefs. But which aspects of linguistic communication are critical? Lohmann and Tomasello (2003) gave three-year-old children who had failed a false belief task three sessions of training and then readministered a similar but different false belief task. The key finding was that participating in so-called perspective-shifting discourse about deceptive objects—for example, discussing the fact that an object looked like one thing but was really another—led these youngsters to an understanding of false beliefs (whereas experience in a control condition did not). It was further helpful if the training included propositional attitude constructions such as "He knows that it's an eraser" or "He believes that it's a cat," presumably because such constructions encode a process of perspective-shifting within a single sentence: the main clause "he knows" or "he believes" signals the possibility of different perspectives or beliefs about the fact that the object is an eraser or cat.

Why does discourse of this type lead three-year-olds to an understanding of false beliefs? Children have nonlinguistic experience all day every day in which they believe something to be the case and then it turns out not to be, or in which they see a person making a mistake that he would never make if he understood the objective situation. Why are such individual nonlinguistic experiences not enough, as they were not in the control condition of the Lohmann and Tomasello study? O'Madagain and Tomasello (2021) argue that propositions expressed in a conventional language are public expressions of the speaker's mental content, and this public expression makes the mental content available as a focus of joint attention with others. And with joint attention to mental content comes the possibility of different perspectives that are simultaneously present but conflicting in some way. When the linguistic expression is a truth-bearing proposition, different discourse perspectives on its mental (propositional) content can actually conflict in the sense that both cannot correspond in a straightforward way to the objective situation. So the reason that perspective-shifting discourse (joint attention to mental content) is effective in facilitating an understanding of false belief is that it facilitates children's coordination of perspectives around their newly emerging sense of the objective situation.

Second, the other variable consistently found to correlate with false belief understanding is executive function. Many studies have found relatively strong correlations of false belief understanding with one or another skill of executive function, and, moreover, longitudinal studies suggest that it is executive function that facilitates false belief understanding and not the other way around (see Devine & Hughes, 2014, for a review and meta-analysis). Recently, experimental methods ("depleting" executive function skills by having the individual work concurrently on a demanding executive task) provide further evidence of a causal link going from executive function to false belief understanding and not the other way around (Powell & Carey, 2017). But again, there is no consensus about precisely which skills of executive function are responsible.

The most common interpretation is that false belief understanding requires the child to inhibit her own understanding of the objective situation. However, three different studies suggest that more important are metacognitive skills for coordinating perspectives or mental states. First, in the meta-analysis of Devine and Hughes (2014; see also Diaz & Farrar, 2017), the strongest correlation with false belief understanding across many studies was not any measure of delay of gratification (inhibition only) but rather the Dimension Change Card Sort task, which measures something more like coordination of perspectives. Further, Fizke et al. (2014) administered several measures of executive function and several measures of mental state understanding to four-year-olds and also concluded that the key is coordination of perspectives, especially "coordinating others' and one's own conflicting perspectives." In addition, recent studies have found that: (i) performance on implicit false belief tasks does not correlate with executive function (perhaps because these tasks do not require coordination of perspectives), whereas classic false belief tasks do correlate with executive function (perhaps because they do involve coordination of perspectives; Grosse-Wiesman et al., 2017); (ii) toddlers' skills of coordinating joint attention predict their later skills of false belief understanding longitudinally (e.g., Sodian & Kristen-Antonow, 2015); and (iii) the coordinating joint attention skills of toddlers with autism correlate with their later skills of false belief understanding (Charman et al., 2000). It would thus seem that the key executive skills predicting children's false belief understanding are not those involving inhibition, but rather those involving the coordination of perspectives (as in joint attention).

For children to acquire an understanding of false belief, then, they must develop some notion of an objective situation against which different subjective perspectives may be evaluated for accuracy, a notion that becomes available only after three years of age with the emergence of skills of collective intentionality. This process requires the mental coordination of simultaneously available conceptual perspectives, which takes place most readily in perspective-shifting discourse (especially when it uses propositional attitude constructions). Such mental coordination of conceptual perspectives requires a metacognitive tier of functioning, with skills of metacognitive re-representation then enabling the construction of a variety of new "objective" concepts from this mental coordination.

7.2.7. Summary

My proposal is thus that the transition to early childhood at three to four years of age is when young children first acquire the ability to construct knowledge that serves to coordinate conceptual perspectives simultaneously and to synthesize them into a single concept, in some cases involving a universalizing objective perspective. What underlies this new ability is the emergence of the metacognitive tier of functioning along with skills of collective intentionality. These twin developments set the stage for processes of constructive thinking and metacognitive re-representation to operate on the mental coordinations and perspective-taking in which children participate, and to inject into the process an objective perspective. Importantly, Rakoczy (2017, 2022) has provided both empirical evidence and theoretical interpretations suggesting that all of the perspective problems, including false belief, are correlated with one another, suggesting some common underlying cognitive processes.

Great apes do not understand conceptual perspectives because, even though they have the possibility of metacognitive reflection, the material on which they are reflecting is not perspectival in the first place. And even though toddlers before three years of age are working with attentional perspectives, they do not have the skills of metacognition enabling them to understand conceptual perspectives, nor the skills of metacognition and collective intentionality that would enable them to coordinate those perspectives with an objective perspective. But after three to four years of age preschoolers (i) coordinate conceptual perspectives with others in shared agencies, (ii) understand an objective perspective, and, perhaps with the aid of linguistic symbols, (iii) metacognitively reflect on these processes and so re-represent them as new types of concepts. Children who understand conceptual perspectives will be able to acquire—through their social and linguistic interactions with others—the appropriate use of a whole host of new ways of thinking and talking "objectively."

The "objective" concepts considered in this section form the foundation of mature scientific thinking in that they distinguish (i) subjective from objective **Perspectives**, (ii) **Appearance** from **Reality**, (iii) true from false **Beliefs**, and (iv) valid from invalid **Reasons** for beliefs. Before making these distinctions, two-year-old toddlers' knowledge and theories cannot go very deep in explaining observations in terms of underlying causal principles and processes. In contrast, preschool youngsters' understanding of objectivity paves the way for a more adult-like approach to the scientific process in which different hypotheses and theories are seen as different conceptual perspectives on one and the same objective reality.

7.3. Normative Attitudes and Concepts

Philosophers since Socrates have wondered not only about the world as it objectively is, but also about the world as it normatively ought to be. Two-year-old toddlers are constantly hearing imperatives about what an adult wishes them to do, but it is arguably not until three years of age that they fully comprehend statements about what one "should" or "must" or "ought to" do. Normative statements of this kind are special because they carry with them a supra-individual force, as they purport to take precedence over individuals' personal goals and preferences, as well as a supra-individual generality, even universality, as they purport to apply not just to specific individuals but to persons in general (within some parameters).

Tomasello and Gonzales-Cabrera (in press) argue that normative thinking and attitudes represent uniquely human psychological adaptations for aligning, coordinating, and negotiating one's actions, thoughts, and attitudes with others within shared agencies. Within shared agencies, individuals co-operate with others interdependently with shared goals and values, but at the same time have their own individual preferences and values that must be co-ordinated or negotiated to maintain the co-operation. Infants already express preferences for individuals who behave prosocially (e.g., Hamlin et al., 2007), and toddlers already behave prosocially themselves (Warneken & Tomasello, 2007). But prosocialty is not the same thing as normative morality. Normative morality requires a shared (even universal) ideal standard coming from a larger collectivity, and indeed moral judgments are quite often experienced as indicating the objectively right and wrong way to do things.

The first shared ideal standards, as argued in Chapter 5, are those that toddlers understand as they collaborate toward a joint goal: what any individual in a particular role must do to ensure joint success. But these are understood only instrumentally in the context of particular activities. When a partner plays her role poorly, toddlers simply exhort her to do something different. But after the "normative turn" at three years of age, preschool youngsters begin dealing with such situations by telling their partner what she "must," "should," or "ought" to do (Tomasello, 2018b). Here I review four sets of normative concepts, two concerning joint collaborative activities (commitment, fairness) and two concerning the cultural group (social norms, institutional facts). Again, I am not aiming at a full explanation but only at establishing that (i) these concepts emerge only after three years of age, and (ii) key to their acquisition is the metacognitive coordination of different preferences or values, one of which is a collectively ideal value.

7.3.1. Commitment/Obligation

When children are competing or acting in parallel with other children, they can ignore their goals and preferences as they wish. But when children are *co*-operating interdependently with partners in joint agencies, they must respect and negotiate with a coequal partner, since both of them have the power to defect or opt out of the cooperation at any point for any reason. Tomasello (2020b) argues and presents evidence that it is only within such interdependent joint agencies that young children begin to express normative attitudes about what one should or ought to do.

When a two-year-old toddler collaborates, it is not clear that she feels a commitment to her partner: if she feels like opting out, she simply does so. But from soon after the third birthday, preschool children seem to feel a commitment to their partner in a joint agency and to expect the same commitment in return. This comes out especially clearly in situations in which young children are explicitly invited to play a game with an adult or a peer ("Let's play X, OK?") and they explicitly accept the invitation ("OK"), that is, the two of them make a joint commitment. When the partner then fails to perform his role as they both know he should, three-year-olds, but not two-year-olds, protest (Kachel et al., 2018). Crucially, this protest is not expressed as a personal preference or desire ("Don't do Y") but rather as a normative requirement ("You must not do Y; you should do X"). The normative requirement in such formulations does not emanate from the individual and her personal preferences but rather from the shared agency and the commitments by means of which it is constituted. Darwall (2006) characterizes joint commitments as partners giving one another the "representative authority" to protest one another's uncooperative behavior on behalf of the shared ideals that constitute the joint agency.

The child also feels the supra-individual force of a joint commitment on her own behavior. For example, when a three-year-old makes a joint commitment to cooperate, she resists even strong temptations to defect (i.e., more often than when she is just playing with a partner in parallel; Gräfenhain et al., 2013; Kachel & Tomasello, 2019). Moreover, when a three-year-old violates a joint commitment, she feels the need to give reasons, excuses, justifications, and apologies to her partner. For instance, when a three-year-old forms a joint commitment with a partner to play a game, and she is successfully lured away to a more fun game, she often acknowledges her breaking of the commitment by asking permission (e.g., "I'll go over there now, OK?") or announcing her departure (e.g., "I'm going over there now"). She does not do this if she is simply playing with the adult in parallel and then is lured away, and two-year-olds do not engage in any of this exculpatory behavior at all (Gräfenhain et al., 2009). Three-year-olds who have broken a joint commitment in the absence of their partner attempt to make amends when he returns (Kanngeisser et al., 2021), or else express their guilt for not living up to their commitment—which, again, two-year-old toddlers do not do (Vaish et al., 2016).

Together these studies suggest that three-year-olds (but not two-year-olds) understand the normative force of joint commitments to shared ideals within (and only within) joint agencies. This is with respect both to their partner's and to their own actions, as evidenced by normative protest from aggrieved parties and by the provision of reasons, justifications, and apologies from aggrievers. They come to this understanding through constructive thinking and metacognitive re-representation of the social negotiations in which they engage as they participate in shared agencies.

7.3.2. Fairness

Young toddlers seem to expect that when people divide resources, they will do so equally among recipients (Geraci & Surian, 2011; Schmidt & Sommerville, 2011). But this is only an expectation, not a normative judgment of fairness.

A normative judgment of fairness is a judgment about what the different recipients deserve, which implies some kind of normative standard.

Once again in the domain of fairness, there is a kind of normative turn at around three years of age, and this is clearest in shared agencies. Three-year-olds do not typically divide resources fairly when they are simply given resources and told that they can share them with others as they like (i.e., in a dictator game; see Ibbotson, 2014, for a review). But when children are collaborating with a partner in a shared agency, the sense of fairness in dividing collaboratively produced resources suddenly becomes salient, perhaps because collaboration activates a sense of self-other equivalence among partners, as argued in Chapter 5, and neither partner "owns" the resources. Thus, Warneken et al. (2011) found that in the context of a collaborative activity, three-year-old children mostly divided the spoils equally, and if one child attempted to take more than half the rewards, she was met with normative protest, and the greedy child almost always relented—presumably because she knew that her greediness was not fair to her coequal partner.

Hamann et al. [2011] added a twist to this paradigm. Pairs of two- and three-year-olds always ended up in a situation in which one of them had three rewards (the lucky child) and the other had only one reward (the unlucky child) so that to create an equal distribution the lucky child would have to sacrifice. What differed across conditions was what led to the asymmetrical distribution. In one condition, the unequal distribution resulted from participants simply walking into the room and finding three versus one reward at each end of a platform. In this case, the lucky child almost never shared with the partner. But in another condition, the asymmetrical rewards resulted from a collaborative effort on the part of the two children pulling together. In this case, lucky three-year-olds, but not two-year-olds, shared with the unlucky child (to create an equal 2:2 split) almost every time. Presumably, they felt that if they both worked equally to produce the rewards, then they both deserved them equally (see also Ng et al., 2011). When, in the same situation, one child works harder than the other, three-year-olds compute deservingness by taking into account their unequal efforts (Hamann et al., 2014).

According to Engelmann and)Tomasello (2019, treating others fairly means treating them with the respect they deserve. In the context of mutualistic collaboration, we both worked equally, and so taking more than half shows disrespect for the partner and his efforts. This is why collaborating children are not just disappointed to receive less than their partner, but positively resentful, and this shows up clearly in their normative protest: they deserve to be treated with equal respect. Importantly, children also do not consider it fair if they receive more than their partner—because they genuinely see her as deserving of equal respect. In general, collaborators do not just prefer that we share the spoils equally, but they feel a normative sense that we owe it to one another as mutually respectful partners. The central role of such "recognition respect" (as the philosophers call it) comes out especially clearly in so-called procedural fairness. Thus, when a child in a collaborative activity gets less than an equal share of the spoils, she is nevertheless happy with that and does not protest *if* (and only if) she had an equal chance at those resources, for example, the resources were distributed by the rolling of dice (Grocke et al., 2015; Shaw et al., 2014). The unlucky child in such cases is not resentful because she feels she was treated with the respect she deserves.

Again, the conclusion is that three-year-olds (but not two-year-olds) understand a normative sense of fairness in dividing resources within (and only within) joint agencies—based on a sense of mutual respect for the ideal way that one should treat one's coequal partner. Again, they come to this understanding through constructive thinking and metacognitive re-representation of the social negotiations in which they engage as they experience resource distributions in social situations.

7.3.3. Social Norms

Human infants are born into the ideal values of their culture as embodied in the group-level norms and expectations that govern individuals' behavior. But infants and toddlers do not experience social norms in this way. When adults attempt to regulate a two-year-old's behavior by prescribing and proscribing certain actions, she understands these attempts only as the exhortations of particular adults. Children begin to understand at least some exhortations as representing the group's common ground expectations for individual behavior (even though they are enforced by individuals) only at around three years of age.

The best evidence for this new understanding is children's own attempts, starting at around three years of age, to enforce social norms on others. For example, Vaish et al. (2011) found that if a puppet begins to destroy someone else's property, three-year-olds will intervene to stop the transgression. Because the child herself is not being affected, she is not protesting how "you" are treating "me." What she is protesting is a lack of conformity to the group-minded social ideal for how one ought to treat others. This interpretation is bolstered by the observation that young children also intervene against individuals who violate mere conventions. Thus, Rakoczy et al. (2008, 2010) found that if three-year-olds (but not two-year-olds) learn that on this table we play the game this way (while on another table we play it differently), and then a puppet plays the game the wrong way for this table, children intervene and stop him, even though no harm is being done to anyone. The child is not defending either her own or any other individual's self-interest; the immediate goal is simply for the wayward actor to conform to the ideal way of doing things.

Preschool children enforce social norms using normative language (Köymen, et al., 2014), suggesting that the enforcer is not just acting as an individual expressing a personal opinion, but rather as a kind of representative of the cultural group conveying impartial and objective knowledge about how "we" act. Norm enforcers are thus, in effect, referring the violator to an objective world of ideal values that he himself may consult to see that his behavior is wrong. Preschoolers' understanding of this comes out especially clearly when they make up their own rules. For example, recent studies have observed children in situations in which there are no established norms or rules, and so, for social order, they invent some for themselves (e.g., Hardecker et al., 2017). The general finding is that five-year-olds make up their own rules and then normatively enforce them on others, suggesting an understanding that self-created rules are as authoritative as any other, presumably because they are based on agreements among coequal partners. Three-year-olds have never been observed to create with peers a social norm or rule, but Schmidt et al. (2016a) found that when a puppet violated a norm, three-year-olds enforced that norm on the violator only if he had actually entered into the agreement.

Importantly, all social norms have a universal dimension in the sense that they apply to *anyone* who is in a certain group or who engages in a certain activity (e.g., anyone who comes into this classroom must hang her coat here)—assuming that they share the norm in common ground with relevant others. Not accidentally, social norms are expressed in the same generic language used in pedagogy as reflecting supra-individual, perhaps even universal, values within some group or context. This metacognitive, collective way of looking at things underlies children's construction—via processes of constructive thinking and metacognitive re-representation—of all multi-perspectival concepts that rely on a universal value as normative ideal.

7.3.4. Institutional Facts

In one of the most curious phenomena of the natural world, individuals extend notions of objectivity and normativity to their social-institutional worlds and thereby create what Searle (1995) calls social or institutional facts. Social or institutional facts comprise real and powerful entities such as: husbands, wives, and parents and their respective rights and responsibilities (created by collective recognition of the cultural ritual of a marriage ceremony); leaders or chiefs and their rights and responsibilities (created by group consensus and sometimes a ceremony); lawyers and doctors and their rights and responsibilities; and so forth. They also can turn otherwise ordinary objects, such as shells or pieces of paper, into culturally potent entities such as money. The phenomenon is that a normal person or object acquires a new status based solely on the deontic powers she is collectively given by the group via some form of agreement, and that agreement is objectified and so becomes part of external reality. Clever as they are, chimpanzees (and human infants and toddlers) cannot act meaningfully in modern humans' social-institutional world—they do not recognize chiefs and money with their respective deontic powers—because they do not have the capability of conferring new normative statuses on otherwise ordinary persons and objects by collective agreement and recognition.

Searle (1995) has been most explicit about how this process works. First, obviously, is some kind of mutual agreement or joint acceptance among group members to designate, for example, an individual as chief. Second, there must be some kind of symbolizing capacity so as to enact Searle's well-known formula "X counts as Y in context C" (X counts as chief in the context of group decision-making). Rakoczy and Tomasello (2007) argue that the ontogenetic cradle of such cultural status functions is young children's joint pretense when they, for example, designate together a stick to be a horse. In doing this they are engaging in the fundamental act that creates new statuses since this designation is a socially public agreement with one's play partner. Importantly, although the ability to pretend derives from very basic abilities of imaginative representation in toddlerhood, the normative dimension comes only with the group-mindedness and collectivity characteristic of three-year-olds. Thus, Wyman et al. (2009) found that once three-year-olds had designated the pretense status of an object with an adult (e.g., this stick is a horse), they objected normatively if someone then treated it as something else.

It turns out that understanding the institutional realities of a modern culture is something that it takes children many years to do. Thus, when they must distinguish institutional objects (e.g., money, drivers' licenses, and borders) from everyday artifacts (e.g., hammers, chairs, and cars)—based on some notion of collective agreement and acceptance—children are not sure of themselves until the school years (Noyes et al., 2018; see also Noyes et al., 2020, for similar findings regarding young children's ability to distinguish the actions that can only be performed by individuals in certain institutional roles [e.g., the president declaring war, an umpire declaring a player "out"] from everyday actions). In any case, the main point in the current context is that children only begin down this road of institutional realities with a joint pretense at three years of age, and they cannot use this new ability to comprehend the complex institutional realities of the modern world until some years later.

7.3.5. Developmental Explanations

Once again, I do not claim to have provided anything close to a full analysis or explanation of normatively structured sociomoral knowledge and concepts. What I have attempted to establish is only three things. First, for almost all of the normative knowledge and concepts I have considered, age three is key. In every experiment in which two-year-olds were tested and compared to three-year-olds, the contrast is stark. This is true of everything from joint commitments to guilt to fairness to cultural common ground to moral justifications. Second, all of this knowledge and these concepts are multi-perspectival, that is, they involve different perspectives or values on the same situation and, whenever individuals act together in collaboration or in the context of a cultural group, this requires some kind of coordination or negotiation of these perspectives and values, including giving reasons and justifications for one's own preferences and values. And third, what defines the knowledge and concepts we have considered here—their normativity—is that within the mix of preferences/goals/values involved, there is a collectively ideal norm that is experienced as a kind of objective value. The ultimate source of these normative ideals, it would seem, is the cultural group "we," whose norms and values supersede those of the individuals who identify with it. It is likely that this kind of normative psychology emerges naturally at age three as part of group-minded collective intentionality, but the particular normative ideals with which young children work clearly must be learned from their sociocultural experiences.

Despite operating with a metacognitive tier of regulation, great apes do not have normative attitudes because they do not operate with collective intentionality, and toddlers before three years of age do not have normative attitudes because they do not operate with either metacognition or collective intentionality. In contrast, preschoolers after three years of age coordinate their preferences and attitudes with others in various shared agencies and metacognitively reflect on this process and thereby recognize collectively ideal normative attitudes around which they can construct various specific normative concepts. Preschool youngsters who understand normative attitudes are now equipped to engage in some new ways of thinking and talking, for example, in commenting that someone "should," "must," or "ought" to do something, or that something is "fair" or "unfair" or the "right" or "wrong" thing to do. There is a normative turn at around three years of age.

A fuller understanding of the ontogenetic emergence of normative attitudes and concepts awaits studies of the type that have been conducted with respect to false beliefs. That is to say, there are few if any studies connecting the emergence of sociomoral normativity to such things as perspective-shifting discourse or cognitive flexibility or any other form of executive or metacognitive functioning. My prediction is that if such studies were conducted, there would be relationships of the same general type that hold in the domain of false belief understanding, namely, children's emerging understanding of normative attitudes and concepts either requires or is strongly influenced by certain kinds of discourse coordination and negotiation—perhaps especially reason-giving discourse in moral discussions—as well as by metacognitive skills for reflecting on these kinds of sociocultural interactions. Kruger (1992) provides some suggestive evidence of this type, but only for school-age children. More apposite here, in a training study Li and Tomasello (2022) found that children who had discourse experiences of disagreement and/or discourse experiences of being asked to justify themselves shifted toward making more flexible and equitable moral decisions based on common ground norms and values.

7.3.6. Summary

The knowledge and concepts we have examined in this section enable preschool youngsters (but not toddlers) to make normative judgments in various kinds of sociomoral interactions, which are essential for their development as competent members of a cultural group. In addition, and continuing with the theme of child as scientist, these normative concepts are crucial in formulating scientific descriptions and explanations of human behavior in all the human sciences. It would be basically impossible to understand what is studied in sociology, social psychology, (behavioral) economics, (cultural) anthropology, history, and political science without understanding instantiations of normative concepts such as **Commitment/Obligation, Fairness, Social Norm,** and **Institutional Fact,** which provide the glue that binds human individuals to one another and to cultural collectivities. Once again, then, the knowledge and concepts that preschool youngsters are beginning to master—based on an understanding of normativity enabled by metacognitive functioning and skills of collective intentionality—constitute a key transition in the passage from the immature social-scientific thinking of toddlers to the more mature social-scientific thinking of adult scientists.

7.4. Real and Ideal

In their most recent theoretical formulations, Carey (2009, 2011) and Spelke (2022) both begin their accounts by noting that humans seem to operate with some species-unique concepts, which any theory of human cognitive development should be capable of explaining. The ones they propose as examples are: *evolution, electron, cancer, infinity, galaxy, tort, entropy, Abelian group, mannerism, icon, deconstruction, wisdom, kayak, fraction, gene, circle, six, wish,* and *good*. Spelke refers to such concepts as "abstract," and Carey (2009) refers to them as providing a "rich conceptual understanding of the world." While providing no specific examples, Gopnik and Wellman (2012) characterize mature human concepts as "coherent, abstract, and highly structured representations of the world," and Xu (2019) characterizes them as "structured, abstract representations that are theory-like."

I do not dispute that these concepts are abstract and complex and, for the most part, uniquely human; but most of them are not universally human. Some of them emanate from scientific institutions that exist in only some cultures (e.g., *electron, Abelian group*), and others are culturally specific because only some cultures engage in the relevant practices (e.g., *kayak, six*). *Good* (in the sense it is intended) is very likely both uniquely human and universal human, but its meaning undoubtedly differs widely across cultures. I would therefore like to propose another list—another kind of list, that is—of unique and universal human conceptual representations. It is a different kind of list because it focuses on representations that emerge, for the most part, during early childhood and, for the most part, across all human cultures. Moreover, my explanation for why they are uniquely human does not rest on their abstractness, structuredness, or richness; I believe that the conceptual representations of great apes are also abstract, structured, and rich in many of the same ways. It rather focuses on how they coordinate and regulate different perspectives.

The knowledge and concepts on which I have focused in this chapter are a special subset of multi-perspectival conceptualizations that incorporate a collectively universal ideal as one crucial perspective in the mix. They thus involve the coordination of perspectives and/or preferences, as do multi-perspectival concepts in general. But the emergence of universalized collective perspectives enables the metacognitive conceptualization of some universalized ideals around which to organize these multiple perspectives. Great apes cannot construct such concepts because they lack collective (or any other shared) intentionality, and toddlers cannot construct such concepts because they lack metacognitive capacities altogether. The universalized ideals that make these types of concepts possible are "objective" and/or normative, and as such they usher the child into yet another existential realm: the realm of normative necessities.

Figure 7.1 provides a graphic depiction of three of the concepts from this chapter. Although these diagrams are highly oversimplified, they are intended simply to make the point that objective and normative concepts require a coordination of social perspectives and attitudes around a collective ideal.

Modern cognitive-developmental theory does not provide an explanation for why these types of concepts emerge only after three years of age. Perhaps because of a rejection of stage theories in general, or perhaps because of an attachment to the notion of domain specificity, modern theories do not look for across-the-board developmental patterns. My explanation for this age of emergence for these types of concepts rests primarily on the hypothesis of an architectural reorganization that occurs at three years of age. This architectural reorganization does not determine the particular content of any of the knowledge or concepts that emerge at this age; it only creates the possibility that certain types of concepts might now be constructed if the child has the right kinds of experiences, including most especially sociocultural experiences involving language and pedagogy. And such construction also depends on certain kinds of prerequisite knowledge and concepts—some already constructed during toddlerhood—involving roles, perspectival representations, and

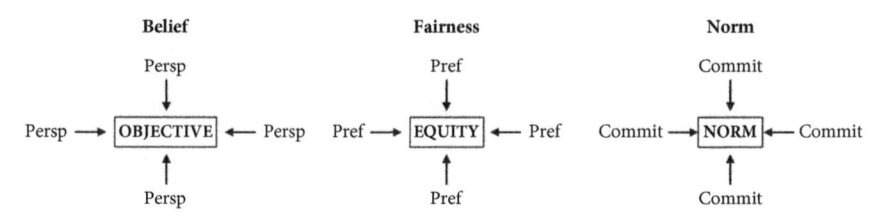

Figure 7.1. Highly oversimplified graphic depictions of three objective/normative concepts from early childhood. Arrows depict different perspectives or preferences or commitments; the box indicates objective/normative ideal. Persp = perspective; Pref = preference; Commit = commitment.

the recognition of self-other equivalence, both epistemically and morally. Preschoolers' architectural reorganization—along with these prerequisite processes—thus enables new processes of metacognitive re-representation involving the coordination and synthesis of perspectives, some of which (due to the skills of collective intentionality that go along with this architectural reorganization) include universalized epistemic or sociomoral ideals. This new way of operating enables preschool youngsters to regulate their thoughts and actions via a kind of we > me normative regulation.

My view is thus that young children do not come to distinguish between subjective perspectives and the objective situation, or to contrast actual behavior with the ideal behavior prescribed by moral principles and norms, by making individual judgments. They come to this way of operating by reflecting on processes of social coordination and negotiation in which there is a triangulation between the simultaneously available perspectives and preferences of individuals, who regard one another's perspectives and preferences as equivalent to their own, in the context of some collectively accepted objective/normative ideals. This happens m ost naturally when individuals are coordinating or negotiating within a shared agency—perhaps especially in the medium of a public language—as this supplies the raw materials with which metacognitive re-representation may construct objective and normative concepts. Objective and normative concepts transition the youngster from the world of possibilities into the world of objective and normative **necessities.**

IV
MOVING FORWARD

Though the world does not change with a change of paradigm, the scientist afterward works in a different world.
                                            Thomas Kuhn

8
An Agency-Based Model of Human Cognitive Development

Most research in modern cognitive-developmental psychology focuses on children's acquisition of knowledge and concepts in specific content domains—from space to objects to mathematics to psychology—and their mastery of specific cognitive skills such as tool use, linguistic communication, collaboration, and reputation management. Each of these knowledge domains and cognitive skills has its own unique properties and so must be studied on its own terms. This is a key insight of the modern theory of cognitive development.

The central argument of the preceding seven chapters of this volume, however, is that all this knowledge and all these skills exist within a domain-neutral psychological organization evolved to facilitate effective decision-making and action, that is, within an agency-based control system architecture comprising goals and intentions, perception and attention, cognitive representations and operations, decision-making and action, executive regulation and learning. The developmental proposal is that human cognitive ontogeny occurs within a series of qualitatively distinct architectures, each conserved from one of humans' ancient evolutionary ancestors as an adaptation for a particular type of agentive decision-making in the context of a particular type of ecological unpredictability. Young infants, toddlers, and preschoolers thus operate within different agentive architectures—toddlers and preschoolers within both individual and shared versions—and these structure the ways in which children of each age experience and learn about the world. To repeat: developing children are not best characterized as Bayesian learners but rather, more generally, as Bayesian agents who learn in support of their agency.

Since the main aim of cognitive-developmental theories is to describe and explain human cognitive development, I wrap things up in this final substantive chapter by summing up my agency-based model's manner of describing and explaining things. In section 8.1, I focus on description, specifically, on the two key transitions in early cognitive development: the 9–12-month transition and the 3–4-year transition. In section 8.2, I focus on explanation, that is,

Agency and Cognitive Development. Michael Tomasello, Oxford University Press. © Michael Tomasello 2024.
DOI: 10.1093/9780191998294.003.0008

on the mechanisms of developmental change that account both for these two transitions and for more local developmental progress. Figure 8.1 presents a schematized summary description of the proposed steps in human agency and cognitive ontogeny and the major processes of learning and re-representation that explain developmental change. This figure will serve as a kind of road map for the discussion.

8.1. Two Key Transitions in Human Cognitive Development

Everyone recognizes that the concepts involved in learning about natural numbers are different from the concepts involved in learning about social norms. The concepts are different because the human experiences on which they are based are different. Everyone also recognizes that there are some psychological processes that operate across domains, from intentions to attention to decision-making to cognitive representation to executive processes to learning. Although these domain-neutral psychological processes may manifest somewhat differently with different cognitive content, from a functional point of view they operate in similar ways no matter the content. That is because cognitive processes operate within an overall psychological architecture evolved for the single function of making cognitively informed behavioral decisions. Agentive architecture thus provides organizational coherence and direction to human cognition across domains, which is why developmental changes of this architecture play such an important role in determining what children of different ages are capable of experiencing and learning.

8.1.1. The 9–12-Month Transition: Agentive Organization

A couple of decades ago Tomasello et al. (1999) proposed that there was a kind of "nine-month revolution" in young children's cognitive development, comprising the relatively sudden emergence of two sets of social-cognitive capacities: (i) the understanding of others as intentional agents whose gaze may be followed, whose actions on objects may be imitated, and whose attention may be manipulated (skills shared with other great apes); and (ii) the coordination of actions and attention with others in joint intentional collaboration involving joint attention and cooperative communication (skills unique to humans). Based on the analyses of the preceding chapters, I now see this nine-month revolution in understanding and coordinating with social

ARCHITECTURE & DECISION-MAKING	COGNITIVE REPRESENTATIONS	COGNITIVE OPERATIONS	LEARNING & RE-REPRESENTATION
Metacognitive Executive Goal → Att ↓ ↗ Act *Reflective Decisions*	Multi-Perspectival + <u>CI</u>: Objective/Normative **Necessities**	Metacognitive Thinking + <u>CI</u>: Coordinating Decisions & Reasons + <u>CI</u>: Universalizing Inferences	Metacognitive Learning: • Belief Revision • CI: Pedagogical Learning • CI: Collaborative Learning + Metacognitive Re-representation
Executive Goal → Att ↓ ↗ Act *Either/or Decisions*	Imaginative + <u>JI</u>: Perspectival + <u>JI</u>: Symbolic **Possibilities**	Executive Thinking + <u>JI</u>: Coordinating Action & Attention + <u>JI</u>: Recursive Inferences	Hypothesis-Directed Learning: • Causality-based (means-ends) • Intention-based (imitative) + Executive Re-representation
Goal → Att ↓ ↗ Act *Go/no-go Decisions*	Iconic **Actualities**	Simple Inferences	Attention-directed Learning: • Bottom-up • Top-down

Figure 8.1. Summary of the three proposed steps of agentive organization in human ontogeny, along with their most distinctive cognitive capacities. First is the goal-directed agency of infants (bottom); second is the intentional agency of toddlers (middle); and third is the rational/metacognitive agency of preschool youngsters (top). 'JI' designates processes derived from the joint intentionality of toddlers, and 'CI' designates processes derived from the collective intentionality of preschoolers.

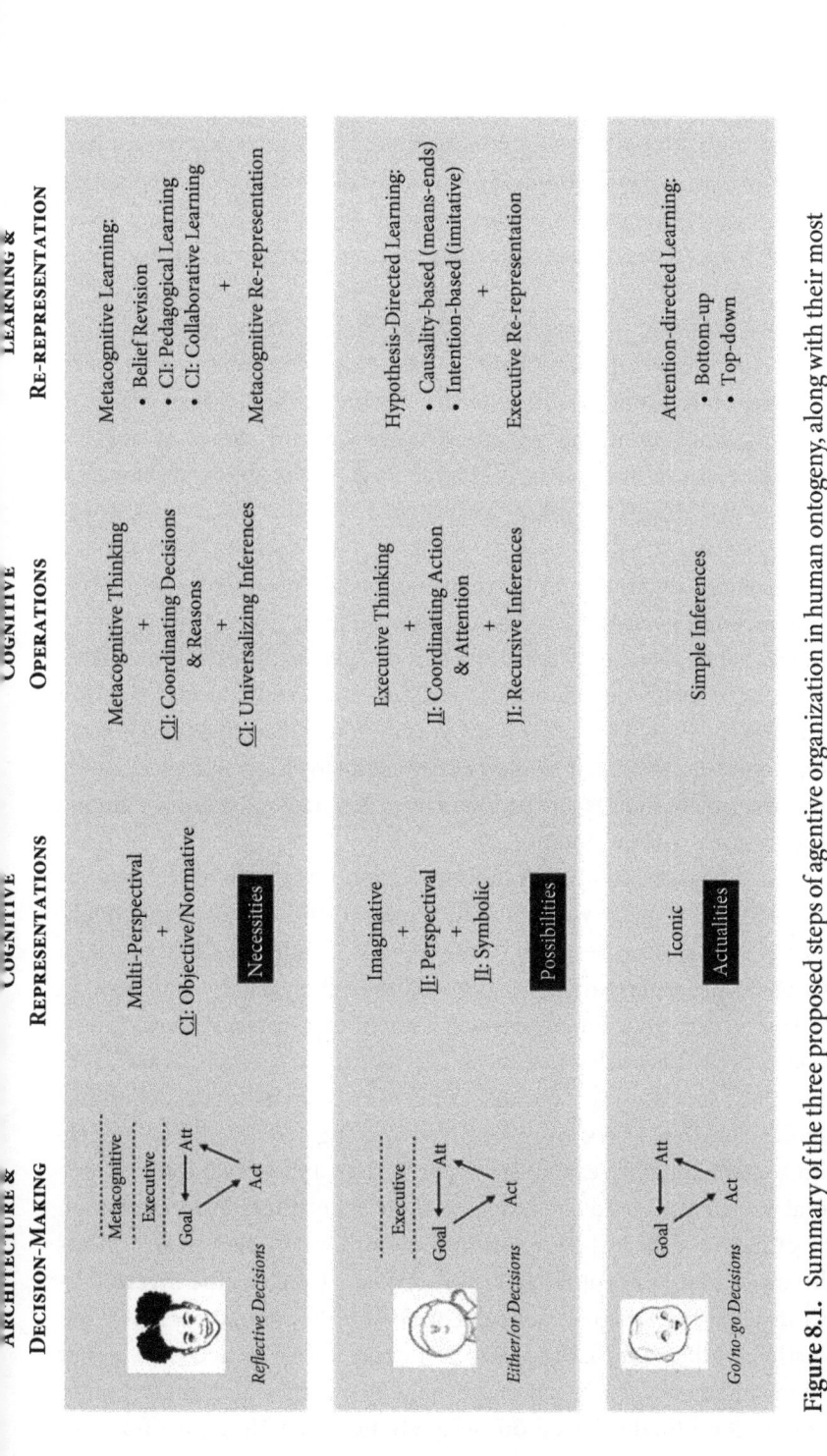

partners as part of a larger developmental transition, specifically, the transition to an architecture of intentional agency. It is still true that cooperating and coordinating with partners in joint agencies is the most distinctive part of this transition since it is the uniquely human part. Nevertheless, explaining why skills of joint agency/intentionality emerge when they do requires reference to the changes in overall agentive organization that occur at around nine months.

With regard to basic processes of decision-making, nine months is the age at which there is a transition from go/no-go decisions about whether to perform an action to more flexible either/or decisions about which action to perform. Thus, eight-month-old infants struggle with the choice of whether to remove one cloth or another to find a toy in A-not-B object permanence tasks, and they also struggle to inhibit prepotent actions in detour tasks. These struggles are consistent with simple go/no-go decision-making being applied in multiple-option situations. Even when young infants select one of two objects that an adult presents to them, it is likely that they are simply attracted to the highest-value object they see. I know of no systematic reports of infants before nine months pausing to look back and forth between options for action or hesitating in the face of difficult choices (as do many mammals and human toddlers in choice situations). It would be very useful to have a systematic study of decision-making in infants and toddlers in which they must choose among possible actions in problem situations.

As intentional agents, toddlers from 9 to 12 months of age begin to make either/or decisions that they executively monitor and control proactively. For example, they choose between removing one or another cloth to find a toy in A-not-B object permanence tasks, and they choose between reaching straightforwardly or circuitously to grasp a toy in detour tasks. Importantly, in both cases they show proactive inhibitory control over a previously successful action by suppressing it before acting. Moreover, in doing such things as using tools, toddlers also show proactive planning in which they organize actions in sequence before executing the first step, and in two different experimental paradigms toddlers monitor their own uncertainty and then make an either/or decision by choosing an "opt-out" alternative. Claims that toddlers cannot conceptualize causal problems such as the inverted Y-tube task in terms of a logical concept of *possibility* may be accurate, but it is nevertheless the case that toddlers can make either/or decisions about their own imagined actions based on probabilistic predictions about potential outcomes. And finally, based on uniquely human adaptations for joint agency, toddlers become capable of forming joint goals and joint attention with a collaborative or communicative partner—if that partner is a supportive adult.

Operating in this way requires imaginative cognitive representations of a type that young infants before nine months of age do not yet possess. Before this age, young infants actively attend to situations that are relevant for them, with relevance determined either by built-in, bottom-up attentional processes—including attention to situations that deviate from expectations—or by more flexible top-down attention aimed at situations relevant for behavioral goals of the moment. Although there are disagreements about their exact nature, most theorists believe that the cognitive representations that guide infants' attention are abstract beyond any concrete sensory content but nevertheless less rich and abstract than the conceptual representations with which older children operate. We may thus characterize infant cognitive representations as perception-based icons (image-schematic). Because they are grounded in goal-relevant attention (either bottom-up or top-down), these representations do not suffer from the indeterminacy characteristic of iconic images (i.e., they are "already interpreted" based on active attention to goal-relevant aspects of the constituent situations). Infants' iconic representations support simple inferences about the world (e.g., that there is an object in a specific state behind an occluder) based on expected relations among objects and events in the actual world.

In contrast, toddlers after nine months of age are able to cognitively simulate or imagine possible actions and their likely outcomes in some possible world. This way of operating enables toddlers to think through possibilities in problem situations before acting and to proactively plan hierarchically organized actions. Many of the imaginative representations that enable this way of operating have the same content as infants' perception-based iconic representations but in a format that enables imaginative manipulation in the context of either/or decision-making. In their joint agencies with others, toddlers understand their partner's role and perspective, which also requires imaginative representations, and in their prelinguistic cooperative communication, they direct others' attention to quasi-propositional situations. Finally, in their linguistic communication toddlers acquire and use linguistic symbols that provide a symbolic format for perspectival representations (e.g., *chase* vs. *flee*, *give* vs. *take*, *I* vs. *you*), as well as grammatical constructions that provide a symbolic format for perspectivizing role-based schemas representing the relational-thematic-narrative dimension of human experience. In both prelinguistic cooperative communication and linguistic communication toddlers make species-unique socially recursive inferences of the form: she intends that I know X.

Why do all of these particular cognitive capacities emerge together? For me, the way that toddlers after nine months of age cognitively represent the world and cognitively operate on these representations via acts of thinking, planning, and either/or decision-making suggests the emergence of a new organizational architecture. Beyond the control system architecture of infants' goal-directed agency, the imaginative and perspectival representations of toddlers' intentional agency can only function on an executive tier of operation. And the recursive inferences required for coordinating action and attention with social partners in joint agencies require some kind of executive workspace on that tier as well. And so a broader view of the nine-month revolution is that it signals the emergence of an executive tier of operation that creates the capacity both for new forms of cognitive representation, inferencing, and decision-making, as well as for joint intentional interactions with others in joint agencies. This new architecture also supports the acquisition of some new dimensions of cognitive content.

8.1.2. The 9–12-Month Transition: New Dimensions of Cognitive Content

Although my focus here has not been on cognitive content, this agency-based model does specify some new dimensions of experience that emerge as cognitive architectures change. According to Spelke (2022), initially in infancy cognitive content is structured by one or another of humans' six core knowledge systems (space, number, etc.), which are domain-specific and modular. If we confine infancy to the first nine months of life, I will not quibble with this description. But from soon after nine months of age toddlers have available to them three important dimensions of experience that are not domain-specific or modular in this same way.

First is the understanding of causality, which is not a core knowledge system precisely because it is not modular but rather applies widely across all kinds of physical events. If we adopt a generally Bayesian approach to children's learning and cognitive development, causality is foundational because it provides the basis not just for recognizing and learning about objects and events in the world but also for formulating and testing hypotheses about *why* things are the way they are. I am proposing, following many others, that the understanding of causality in the outside world is grounded in infants' and toddlers' experience of their own agency in making things happen. Before nine months, young infants experience a contingency between their own actions and outcomes in the world—and they experience external events in this same way—but it is only with the emergence of intentionally structured agency in toddlerhood (in which the only goal of an embedded act is the larger act) that they understand that their agentive action can "force" an event to occur. In using and choosing tools (abilities shared with other great apes) one-year-old toddlers extend this way of thinking to understand how certain properties of the tool, as intermediary, contribute to the causal efficacy of the overall agentive action. Perhaps in combination with an understanding of the causal efficacy of other agents' intentional actions, this leads toddlers (and other apes) to attribute some causal powers or forces to external entities independent of their own actions, thus making possible the first causal hypotheses and theories about the world.

Second is the understanding of intentionality, which goes beyond Spelke's core concept of agency in that, analogous to the case of physical causality, it provides a basis for formulating and testing hypotheses about *why* agents act as they do. Here I am proposing once again that the key to this hypothesis formulation and testing is the experience of one's own agency given certain evolved capacities. Most important is the processes of common coding of self and others' actions (evolved to support apes' aligning of their own actions with others' in social learning). The proposal is that common coding is present from birth in all great apes (it supports neonatal mimicking), and it accounts for the synchronous developmental emergence of infants' production of goal-directed acts and comprehension of the goal-directed acts of others. But it is with the emergence in toddlerhood of embedded intentional actions—in which one act is done only so that another may be done—that children ask why someone is doing something and then propose and test relevant hypotheses (e.g., at 12 months in rational imitation tasks). Such hypotheses are formulated in terms of the very same intentional states, such as goals and attention, with which the toddler operates in her own executive decision-making (based on simulation), but they also incorporate a theory-driven causal hypothesis about the particular actions of particular agents in particular circumstances (based on theoretical thinking). Because both processes are needed, I call this the "simulation + theory" theory of children's basic capacities for understanding the intentional states of others.

Important evidence that children's early understanding of causality and intentionality is based in their agentive action is the systematic structure of the inferences they make. In the physical domain, their inferences are structured into quasi-logical paradigms by various combinations of causal conditionals (e.g., if it contains food, shaking it will make noise) and contrary negations (e.g., if there was only silence, not noise, then it contains no food). These causal inferences are not based on propositional logic and certainties, but rather on a Bayesian logic of probabilities. In the social domain, children's inferences are also structured into probabilistic paradigms based on the logic of practical action (e.g., if his goal is obtaining X and he sees it at location Y, he will go to Y—whereas if he has some other goal or sees something other than the object at location Y, he will stay or go elsewhere). Both types of inferential paradigms are based simply on the way the human organism operates, the logic of its action as a control system: if I act, there will be a result; if I do not act, there will be no result; if there was a result, there must have been an action; if there was no result then there must not have been an action (or else an ineffective action). Causality and intentionality structure the way that great apes in general—and humans from the point of toddlerhood onward—attend to the world and understand its workings based on the logic of agentive actions. It is therefore from this point forward that children are operating as scientists.

An important question for this hypothesis is the extent to which young infants (before nine months) already understand causality and intentionality. In the case of causality, infants as young as six months expect launching events to have a certain structure and expect goal-directed actions to have a certain topography. But in both cases the relevant studies use violation of expectation methods. Such methods cannot establish in the case of causality an understanding of causal intervention, the role of force, or counterfactual reasoning, all of which modern theories consider essential for the concept. And in the case of intentionality, those methods cannot establish anything beyond expectations of particular actions, again without considerations of intervention and counterfactual reasoning of the type that toddlers clearly show. To answer basic questions about young infants' capacities, therefore, I believe that new methods are needed. Most promising for infants just before nine months of age might be imitation paradigms, with which they have some competence (e.g., something like a rational imitation paradigm). For even younger infants, it might be revealing to try methods of predictive looking, which have been used to such good effect with older infants. It would also be informative to try to establish more broadly the relation between infants' and toddlers' understanding of their own intentional action and the causal and intentional structure of events in the external world.

The third key dimension of cognitive content generated by the nine-month transition from goal-directed to intentional agency is joint intentionality and its components, which Spelke (2022) attempts to capture by positing a core notion of "social being" and a derived understanding that some cognitive representations are "sharable." But these are only labels for underlying cognitive processes that need further explication. For example, as just noted, joint agency and intentionality introduce into human experience the species-unique notions of role and perspective. These notions enable children to construct a variety of species-unique perspectival and role-based representations—from the concepts of *I* and *you* to the concept of *policeman* to the propositional structuring of role-based grammatical constructions—which provide human cognition with much of its unmatched power and flexibility. Across toddlerhood, joint agency and intentionality also create the sense of self-other equivalence, which plays a crucial role in toddlers' coming to understand the equal validity of their perspectives and preferences and those of social partners. Like causality and intentionality, then, toddlers' joint intentionality (including the notions of role, perspective, and self-other equivalence) also constitutes a dimension of human experience that is not tied to any specific cognitive domain, but that provides raw materials for the construction of many different types of cognitive content.

My hypothesis is that all of these new dimensions of experience emerge with the nine-month revolution at least partly because they all require imaginative representations and an executive workspace, along with the experience of formulating intentionally structured actions and executively supervising their creation in agentive decision-making and inhibitory control. Causal understanding also relies on the emergence of great ape adaptations for tool use; intentional understanding also relies on apes' common coding of actions; and the understanding of roles, perspectives, and self-other equivalence also relies on the emergence of human adaptations for joint intentionality. All happening on an executive tier of operation, these new dimensions of experience thus enable toddlers to operate as hypothesis-testing scientists able to construct new types of knowledge and concepts for coping with the cognitive challenges of their physical and social worlds.

8.1.3. The 3–4-Year Transition: Agentive Organization

A few years ago Tomasello (2018b,2019) proposed that there was a kind of "objective/normative turn" in young children's cognitive development at around three years of age. Once again, the original motivation was to focus on uniquely human social-cognitive capacities, in this case the ability of preschool children to construct objective and normative knowledge and other concepts based on emerging capacities for collective agency/intentionality. But, once again, based on the analyses of the preceding chapters, I now see this turn as part of a larger developmental transition of organizational architecture. It is still true that constructing objective and normative knowledge and concepts is the most distinctive part of this transition since it is the uniquely human part. Nevertheless, explaining why skills of collective agency/intentionality, along with objective and normative concepts, emerge when they do requires reference to changes in overall agentive organization and architecture that occur at around three to four years of age.

As metacognitive agents, both great apes and preschool youngsters make reflective decisions in which they metacognitively monitor and control their executive-tier processes of thinking, planning, and belief formation. Therefore, even after they have formed a belief and made a decision, apes and preschoolers can determine when new information calls that belief or decision into question in different ways, which requires the metacognitive coordination of beliefs and evidence in the process of belief revision. This is different from just monitoring a feeling of uncertainty in decision-making, which is already characteristic of toddlers. To act as metacognitive agents, preschoolers must be able to think about their own thinking as they make reflective decisions, and they must be able to form metacognitive strategies that reflectively consider difficulties in planning (e.g., by devising metacognitive strategies such as mnemonic aids). Moreover, preschoolers are also able to coordinate with partners not just their actions and attention, as toddlers already do with adults, but also their thinking and decision-making even with co-equal peers, using reasons, evidence, and justifications to connect with the beliefs and values they already share in their personal or cultural common ground.

These processes of metacognitive thinking and decision-making are made possible by preschoolers' ability to form metacognitive representations of their own executive-tier cognitive processes. In addition, reflecting on their mental coordinations with others in shared agencies enables preschool youngsters to construct multi-perspectival knowledge and concepts. This occurs most often in linguistic discourse as youngsters jointly attend to the perspectives (mental content) of one another's utterances simultaneously, and jointly assess the validity of assertions based on reasons. The result is knowledge and concepts about such things as natural numbers, which involve simultaneously coordinating perspectives of cardinality and ordinality (not to mention coordinating perspectives within each of these), and causal networks, which require the simultaneous understanding of particular events as both effects of previous causes and causes of subsequent effects. Multi-perspectival concepts are not acquired via individual (non-social) experience—or even via theory-of-mind perspective-taking from a third-party point of view—but rather they are acquired via participatory engagement with others in shared intentional (collaborative and communicative) interactions that require recursive mental coordination and cooperative self-regulation.

The way preschoolers after three years of age cognitively represent the world and cognitively operate on those representations thus requires a new organizational architecture. Beyond toddlers' intentional agency with its executive tier of operation is preschoolers' metacognitive agency and its metacognitive tier of operation. And mentally coordinating thoughts, plans, and decisions with partners—based on conventions, norms, and the culture's other instruments of collective intentionality—requires a metacognitive workspace on that tier as well. A broader view of the three-year objective/normative turn is therefore that it signals the emergence of a metacognitive tier of operation that makes possible new forms of cognitive representation, inferencing, and decision-making, as well as interactions with others based on cultural conventions and norms. This new form of agentive and cognitive organization creates some new and very powerful dimensions of cognitive content.

8.1.4. The 3–4-Year Transition: New Dimensions of Cognitive Content

Again, my focus has not been on the specifics of what preschool children learn in their everyday worlds, that is, on cognitive content. I have rather focused on what preschoolers after three years of age now have available to them in the form of new dimensions of experience that are crucially important in their gradual construction of cognitive content in many different domains of knowledge. The most important of these new dimensions of experience are the universalized or idealized perspectives or values—"what anyone should believe or do"—that enable the construction of three remarkable types of multi-perspectival concepts.

The first type is objective knowledge and concepts or, more accurately, the distinction between individual perspectives and the objective situation. This way of thinking depends both on preschoolers' newfound ability to think about things from multiple conceptual perspectives simultaneously and on an ability to appreciate that one's own perspective and that of others are basically equivalent. This way of thinking thus constitutes the foundation of mature scientific agency in that it distinguishes subjective from objective perspectives, appearance from reality, true from false beliefs, and valid from invalid reasons for beliefs. Although two-year-old toddlers work with causal hypotheses and theories, older preschoolers' mastery of knowledge and concepts based on these distinctions of objectivity leads to a more mature understanding of the process of scientific belief revision based on the dialectic between observation and underlying causal explanation. Rakoczy (2017, 2022) has provided both empirical evidence and theoretical interpretations that all of these distinctions of objectivity come from a common source. Tomasello (2018a) reviews empirical evidence and identifies this common source as perspective-shifting discourse with others, as this provides young children with clashes of conceptual perspectives needing to be resolved, which individuals do via constructive thinking and re-representation of these clashes (see next section). This leads to a whole host of "objective" concepts that presuppose a comparison to some objective standard, including mental state concepts such as *think, know,* and *believe*, as well as concepts such as *true, false, correct,* and *incorrect*, and also a number of pragmatic discourse markers such as "actually," "supposedly," and "in reality."

The second type is normative knowledge and concepts or, more accurately, the distinction between individual preferences and the right thing to do. This way of thinking depends both on preschoolers' newfound ability to think about things from multiple conceptual perspectives simultaneously and to appreciate the notion of self-other equivalence. This way of thinking thus constitutes the foundation of mature sociomoral agency that distinguish individual preferences from socially normative values. Children of this age thus begin to operate with attitudes of commitment, obligation, promise, fairness, and social norm, as well as using appropriately words such as "should," "must," and "ought," as well as making distinctions between "fair" and "unfair" or "right" and "wrong." Tomasello (2020b) reviews evidence that these ways of thinking and talking emerge first and most naturally within children's joint and collective agencies with social partners (e.g., they share fairly only with collaborative partners, not with bystanders). It is possible—although we do not have direct empirical evidence—that the process of constructing these concepts and words is that children reflect on their value-based coordinations and negotiations with collaborative or communicative partners, that is, as they are collaborating with such partners to regulate their shared agencies. Studies investigating this hypothesis—including training studies with discourse negotiation and conflict resolution—would be of immense value in understanding the origins of normative thinking and attitudes (see Li & Tomasello, 2022, for a modest start in this direction).

The third type of concepts based on universalized or idealized perspectives and values is knowledge and concepts of institutional facts, which typically emerge somewhat later than objective and normative concepts, at least partly because they rely on these concepts as prerequisites. Thus, when preschoolers must distinguish institutional objects (e.g., money, marriage licenses, and countries) from everyday artifacts (e.g., bicycles, tables, and lamps)—based on some notion of collective agreement and acceptance—they are not sure of themselves until late preschool, at least partly because these objects become institutional only when the child is able to assign to the various roles in them normative rights and responsibilities from the point of view of the entire cultural collective. Navigating the institutional world is a challenge for children throughout middle childhood and beyond.

All three of these new types of knowledge and concepts involve multiperspectival conceptualizations that also require a kind of collectively universal ideal as a crucial component. My hypothesis is that these new dimensions of cognitive content are part of the 3–4-year objective/normative turn at least partly because they require metacognitive representations and metacognitive operations in a metacognitive workspace. But this new organizational architecture is not by itself sufficient. Also required are skills and concepts of collective intentionality, including the construction of objective and normative ideals. The outcome is that these new dimensions of cognitive content enable preschoolers for the first time to make the kinds of objective and normative judgments—and to construct the kinds of objective and normative concepts—that are essential for full participation in a cultural group.

8.1.5. Three Existential Modalities

The bread and butter of modern cognitive-developmental research is establishing when and how young children acquire particular cognitive content in particular cognitive domains. As for organizational architecture, the modern theory mostly assumes that cognitive representations, operations, and decision-making are of the same general nature throughout the first years of life, with developmental changes simply reflecting quantitative increases in complexity, abstractness, and richness (with perhaps some qualitative differences due to the acquisition of language). In contrast, my proposal is that these quantitative increases in complexity, abstractness, and richness are in many cases the result of qualitative changes in underlying cognitive representations and operations due to changes in overall organizational architecture. Further, modern cognitive-developmental theorists mostly assume that cognitive content is wholly domain-specific. While this is certainly true at some level of analysis, there are also domain-neutral dimensions of human cognitive content—such things as causality, intentionality, role, perspective, objectivity, normativity—that provide essential building blocks for the construction of much of this domain-specific content. Describing a particular cognitive-developmental pathway therefore requires an account not only of the domain-specific aspects of the cognitive content involved but also of the overall agentive architecture and domain-neutral dimensions of experience involved.

Based on this theoretical approach, my general conclusion is that children at different ages live and learn in different experiential worlds—just like scientists at different points in history, as described by Thomas Kuhn in the epigraph to this section. One piece of evidence for this rather radical claim is the fact that the knowledge of young infants, toddlers, and preschooler youngsters is typically assessed with different experimental methods. Since young infants' goal-directed agency uses only perception-based iconic representations that guide attention, attention is what developmentalists set out to measure in the various looking-time paradigms. Since toddlers' intentional agency enables them to imaginatively represent experiences and so to actively use their representations in thinking, planning, and either/or decision-making, these processes are what developmentalists set out to measure in various action-based paradigms. And since preschoolers' metacognitive agency enables them to represent their experiences and mental operations from multiple perspectives simultaneously, developmentalists often ask children of this age to reflect on, or even linguistically express, their knowledge.

And so my most general conclusion is that infants, toddlers, and preschoolers, inhabit different experiential worlds in the sense that they cognitively represent their experience in three different existential modalities (a la Kant, 1781, and modern modal logic). In their goal-directed agency infants represent the world as **actualities** (in perception-based iconic representations); in their intentional and joint agencies toddlers represent the world also as **possibilities** (in imaginative and perspectival representations); and in their metacognitive and collective agencies preschool youngsters represent the world also as **objective and normative necessities** (in objective and normative metarepresentations). The way children's experience is structured by these different representational formats helps to determine when and how they learn particular types of cognitive content.

8.2. Processes of Developmental Change

To explain developmental change and the emergence of new knowledge, concepts, and cognitive organization in children's cognitive development, the two most basic processes are, of course, maturation and learning. In general, maturation builds the organizational and processual framework, and learning fills in cognitive content (although some cognitive content may emerge maturationally, and language learning may contribute to processes of cognitive representation and thinking). But maturation and learning are not the whole story. Also necessary for cognitive-developmental progress are endogenous processes of constructive thinking and re-representation via which children become able both to generate new hypotheses in their Bayesian learning and to reflect internally so as to reconcile and harmonize inconsistencies and inefficiencies in existing knowledge.

8.2.1. Maturation

Maturational processes contribute to human cognitive development in at least three ways. First, some cognitive content emerges in human ontogeny without learning or environmental input, for example, infant core cognition. Second, some domain-neutral cognitive processes mature in ways that have widespread effects, for example, the executive workspace may expand its capacity. And third, the organizational architectures of human agency—including shared agency—mature and usher in new ways of operating, giving rise to new capacities for cognitive construction and learning based on new types of cognitive representations, operations, and self-regulation. In combination with the maturation of other capacities, the maturation of organizational architectures also gives rise to new dimensions of experience that serve as raw material for the construction of new types of cognitive content (e.g., various objective and normative concepts).

In evolution, the most fundamental developmental pathways of a species are normally inherited from ancient ancestors. In the current context, this raises the question of which aspects of human agentive and cognitive organization are conserved from which evolutionary ancestors. Although those evolutionary ancestors are long gone—and their psychology with them—one can examine extant species that are plausible models for these earlier ancestors, including in experiments aimed at assessing their agentive and cognitive functioning.

Tomasello (2022a) performs this kind of analysis and concludes, as noted in Chapter 2, that infants' goal-directed agency was characteristic already of the earliest vertebrates, toddlers' intentional agency was characteristic already of the earliest mammals, and preschoolers' metacognitive agency was characteristic already of the earliest great apes—with toddlers' and preschoolers' processes of joint and collective agency only emerging, respectively, with early and modern humans a few hundred thousand years ago. Figure 8.2 applies this account to human ontogeny, adding in some other agentive and cognitive building blocks and their evolutionary sources, including especially the understanding of causality (associated with the evolutionary emergence of great ape tool use) and intentionality (associated with the evolutionary emergence of great ape common coding in the context of social learning), along with the roles, perspectives, and self-other equivalence characteristic of shared agencies.

We know very little about the cognitive ontogeny of any nonhuman species, and so we can only speculate about how these capacities may have shifted around in developmental timing from earlier to later species. One domain about which we have some knowledge is social cognition. Based on a variety of studies, we know that extremely young great ape infants (i.e., chimpanzees and human infants) process the actions of others and themselves in a common code, as evidenced by neonatal imitation. Their understanding of the intentional actions and states of others then follows in both species, but at a slightly younger age in humans, perhaps as adaptations to cooperative childcare (Tomasello, 2020a). We know next to nothing about the ontogeny of decision-making, executive function, and metacognition in nonhuman great apes. Much more research is needed—with the different species observed in comparable contexts—for any firm conclusions to be drawn about how human ontogeny has come to diverge from that of other great apes.

There is thus no doubt that maturation—as the direct expression of human evolution by means of natural selection—plays a key role in structuring human cognitive development and learning. However, in complex human competencies maturation never determines anything on its own. What matures is only a *capacity*, and the realization of that capacity presupposes a normal human environment, of the type supplied by every human culture, and normal processes of learning in the child. Said another way, what matures is a new way of operating, but to acquire cognitive content or knowledge in a particular domain the child must actually operate in that way in experiencing and learning about the world. If a child were reared outside of a normal human environment, or lacked normal human abilities to learn, many atypical outcomes would be possible. Every complex developmental pathway develops

AN AGENCY-BASED MODEL   171

	Vertebrate	Mammal	Great Ape	Early *Homo*
Preschool Youngster			*Metacognitive Agency*	**Collective Agency:** • objectivity • normativity
Toddler		*Intentional Agency*	Theories: • Causal [tools] • Intentional [common coding]	**Joint Agency:** • roles • perspectives • S = O • recursivity
Infant	*Goal-Directed Agency*	Sociality	Common coding [imitation]	Emotion sharing [cooperation]

Figure 8.2. Summary of the hypothesized evolutionary sources of key components of human agentive and cognitive development (S = O indicates the notion of self-other equivalence). The leftmost diagonal labels indicate the most general level of agentive and cognitive organization.

epigenetically, incorporating either more or less environmental input, and its age of first emergence is determined by many factors. The proposal, then, is not that the competencies in Figure 8.2 are "innate"—as a static category—but rather that they develop naturally and in similar ways in all normal human environments without highly specified learning experiences.

8.2.2. Learning

Learning is not inevitable. One could imagine organisms that make behavioral decisions focused only on the situation at hand, without preserving anything of that experience for the future. But learning is adaptive for future decision-making, and it seemingly does not require much extra machinery—that is, beyond the basic self-regulation of agentive control systems—for the individual to preserve its experiences for future use. And so basically all complex organisms learn from experience.

The gateway to learning is attention. It would not be adaptive to preserve every perceptual experience from every waking moment, and so organisms mostly preserve only what is relevant for them and their agentive actions. This means preserving the experiences structured by attention, either bottom-up attention (as determined by natural selection) or top-down attention (as determined by the goals of the individual agent in its ongoing decision-making and action). And so, in the beginning human infants learn from situations to which they attend either because, bottom-up, these deviate from the expectations inherent in their core iconic representations or else because, top-down, they present relevant opportunities and/or obstacles for goal-directed actions. Relevance-structured attention and learning of this type are sufficient for infants because they only need to make go/no-go decisions, which only require recognizing situations.

Independently locomoting toddlers must make more flexible and thoughtful behavioral decisions, and this requires them to attend to and learn about all kinds of possibilities in the world, including those based on physical causality and intentional agency, as they imaginatively represent them on their newly emerging executive tier. This imaginative way of operating on an executive tier leads toddlers to explore the world curiously in ways not directly relevant to their current actions but rather in ways that might be relevant for future actions. Much of this exploration takes the form of hypothesis testing in which the child's immediate goal is simply to see how things work causally. This is the origin of children's first theories and hypothesis-directed learning. Something analogous happens in interactions with other people, as toddlers form theories and test hypotheses (formulated in the same psychological terms they use in regulating their own action and attention) about why individuals do what they do in particular situations. The understanding of intentional agency creates the possibility for toddlers to engage in new forms of social learning in which they, for example, reproduce only the intended acts of others or those that make sense for them given their circumstances. Toddlers' emerging capacities for joint intentionality enable them also to learn about the roles and perspectives of their interactive partners and to create shared (common ground) experiences with them.

In early childhood, preschool youngsters develop a metacognitive tier of functioning along with capacities for collective agency and intentionality. This manner of functioning enables them to learn in three new ways. First, as individuals they may learn metacognitively, for example, as they reflect on and revise their current beliefs in light of new information in rational ways (belief revision), which is new at this age because it requires metacognitive operations. Second, based on their emerging capacities for collective intentionality and its universalizing inferences, preschoolers also for the first time begin to understand adult pedagogy as communicating culturally "objective" knowledge and beliefs that one ought to adopt. And third, these two new dimensions of agentive functioning together enable youngsters to engage in truly collaborative learning with peers in which they coordinate their perspectives with partners based on reasons and evidence. These kinds of collaborative coordinations constitute the raw material on which preschoolers reflect to engage in new types of constructive thinking and to re-represent their experience in terms of multi-perspectival, objective, and normative concepts. A summary of these various types of learning is in Figure 8.3.

	Individual Learning	*Social/Cultural Learning*
Metacognitive Youngsters	Belief Revision [coordination of beliefs with current evidence]	Pedagogical Learning & Collaborative Learning [conceptual perspectives]
Intentional Toddlers	Hypothesis-Directed (causal) Learning	Imitative Learning of Intended Actions
Goal-Directed Infants	Attention-Directed Learning	Mimicking Actions & Emulating Results

Figure 8.3. Summary of different kinds of individual and social learning at different periods of human development (made possible by different forms of agentive organization).

In general, individual organisms experience the world in ways that their species has evolved to experience it, and learning presupposes structured experience. What I have proposed here is that an important part of this structuring in human ontogeny is maturation of the agentive/cognitive architecture characteristic of the species, including its particular formats of cognitive representation and types of cognitive operations at different developmental periods. Learning thus takes many forms across development. In addition, learning can have many kinds of effects. Thus, Gottlieb (2007) notes that individual experience may influence the ontogenetic expression of some cognitive capacity or skill in different ways, for example: (i) it may simply trigger the emergence of some skill; (ii) it may facilitate the emergence of some cognitive capacity or skill while not being strictly necessary; or (iii) it may be a necessary process for the development of some cognitive capacity or skill. The conclusions I have drawn in the previous text and figure—about the age of emergence of different types of learning—are in most cases not based on systematic developmental data on comparable tasks across ages, and so mapping out the different kinds of learning, and the different kinds of effects they may have in particular cases, should be a major goal of future research in cognitive-developmental psychology.

8.2.3. The Role of Language

Vygotsky (1978) emphasized that children's cognitive development depends on a cultural environment providing them with various kinds of cognitive tools and prostheses to facilitate their thinking and learning, everything from physical tools to systems of symbolic representation to adult-taught cognitive strategies. Although there are differences in which tools and protheses different cultures provide, one species-universal and especially important cultural tool is language.

A language provides developing children with a readymade map of the world. As children attempt to comprehend the adult's referential intention in using a linguistic symbol—including the use of that same symbol in different situations over time—they are led to attend to and categorize the world in ways that they might not otherwise do. Children's initial understanding of the referents of linguistic symbols is often immature, and in attempting to comprehend further uses they are pulled into more adult-like ways of attending to and categorizing the world. Carey (2009) provides an account that attempts to be more explicit about the cognitive processes involved. Focusing mainly on logical and scientific concepts, she emphasizes that children's imperfectly understood linguistic symbols act as "placeholders" that structure and focus their ongoing learning. Her hypothesis of Quinian bootstrapping attempts to explicate the cognitive processes involved as children (re)construct the adult concept for placeholder words. For example, preschoolers learn the concept of natural numbers by learning the number words and then constructing mental models that correspond to adults' subsequent use of these words, relying on cognitive processes such as induction and analogy.

Importantly, in this formulation the important cognitive work is being done by attention, categorization, induction, analogy, and other basic cognitive processes, with language serving only to prompt and direct them. Just as reading a map requires pre-existing skills of spatial cognition used in certain new ways, comprehending a linguistic symbol requires pre-existing cognitive skills used in certain new ways. This is clearly the case in the evolution and history of human cognition. For each and every new linguistic symbol, the first persons in history to use it must have constructed the underlying concept without language. If the first users were able to construct conceptualizations without language, then contemporary children presumably can do so as well in the right circumstances (as some deaf children growing up without a conventional language do in some circumstances: Goldin-Meadow, 2005). None of this is to say that language does not facilitate the process immensely as, again, it often prompts children to look at things in ways they would not otherwise do. In addition, Spelke (2022) emphasizes that linguistic symbols also serve to connect knowledge from different domains as children apply the same word to similar phenomena across domains (e.g., "large" may apply to a person, a space, or a number). The point is only that language does not create new conceptualizations out of nothing.

Some theorists have also argued that the grammatical dimension of linguistic competence is crucial. Most radically, Spelke (2022) focuses on the role of the supposed universal grammar with which infants are born (a la Chomsky). This innate grammar enables them to conceptualize and represent new knowledge without any specific learning experiences by grammatically combining existing concepts. Thus, for example, Spelke (2003, p. 306) claims that "Having learned the meanings of *left*, *blue*, and *thing*, [without further experience, the child] knows the meaning of the expression *left of the blue thing*" Spelke (2022, p. 444) therefore argues that one key to cognitive development after infancy is "the innate rules and representations that underlie the learning and use of any natural language," which are "recursively combinatorial and compositional," with infants' core knowledge systems providing the primitive concepts that serve as the initial raw material.

The problem is that other animal species possess a wide array of creative cognitive capacities without a Chomskian universal grammar. Chimpanzees, for example, understand (i) how objects are categorized and even quantified in small numbers; (ii) basic spatial relations among objects (including, very likely "left of the blue thing"); (iii) causal relations among physical events; and (iv) goals and perceptions of other agents. Obviously "the innate rules and representations that underlie ... natural language" are not necessary for apes' cognitive capacities—not to mention the fact that there is no evidence for, and much evidence against, an innate universal grammar in humans (Tomasello, 2003). An alternative hypothesis is that, as in the case of linguistic symbols, the grammatical constructions of a language direct children's attention and categorization in ways that they might not have directed them on their own. For example, both correlational and experimental studies have found that children's acquisition of propositional attitude constructions (e.g., "I think it is in the blue box") provides them with a representational format that facilitates their coming to understand that others may have false beliefs about propositions (DeVilliers & Pyers, 2002; Lohmann & Tomasello, 2003). But, again, in human evolution and history someone must have been the first to think in these terms without language as a guide, and so it is likely that children already possess key cognitive prerequisites involving mental processes when they are first exposed to propositional attitude constructions and these actually facilitate the acquisition these constructions, which then facilitate an understanding of (false) beliefs.

Finally, in addition to the role of linguistic representation and its combinatorial capacities, the acquisition of language facilitates children's cognitive development in still another way. In addition to its role in representing the world, language also structures human social intercourse as individuals attempt to coordinate their thinking with others. Indeed, for many theorists (e.g., Wittgenstein, 1953; Clark, 1996; Tomasello, 2008) linguistic representations only exist to facilitate communicative interaction: the reason one attempts to make reference to things in the world is to coordinate with the thoughts and actions of others. Such attempts actually begin early in ontogeny with the pointing gesture, in which the toddler and adult attempt to direct one another's attention to situations in their shared environment without any representational symbols at all. Then, in late toddlerhood children begin to engage in conversational discourse in which participants share a focus of attention on a topic of conversation in public symbols and then attempt to manipulate the perspective of their partner on that topic in acts of mental coordination. Such joint attention to mental content thus facilitates children's construction of perspectival and multi-perspectival concepts, as they reflect on and re-represent their linguistically mediated mental coordinations with others.

In the current view, then, acquiring a conventional language in its representational, combinatorial, and coordinative functions plays a crucially important role in children's cognitive development in many ways. Attempts to comprehend linguistic symbols and constructions require children to focus their attention in certain ways and to categorize the world in certain ways as well as to take certain perspectives on things, and this marshals and directs their pre-existing cognitive capacities in more mature and culturally conventional ways. But, to repeat, it does not create new conceptualizations out of nothing.

8.2.4. Constructive Thinking and Re-Representation

As powerful as these various forms of maturation and learning may be, they are not the only mechanisms of cognitive-developmental change. There is another set of mechanisms based on endogenous processes in which the individual reflects on her own psychological processes and constructs new knowledge as a result, what I have called constructive thinking and re-representation. The prototype of the process is Archimedes in his bathtub having a moment of "Eureka!" insight in which he spontaneously connects existing ideas to create some new brilliancy.

The executive and metacognitive tiers in human agentive architecture evolved most directly to facilitate effective decision-making in the moment. They also enable new kinds of learning, as just described, to facilitate effective decision-making in the future. My proposal here is that these new tiers of psychological functioning also enable processes of constructive thinking and re-representation in which the individual reflects on her knowledge either to creatively construct some new hypothesis in the moment, or else to eliminate discrepancies and inefficiencies in her knowledge base in anticipation of future decision-making. The process works by (i) generalizing vertically across instances via abstractions and analogies, and (ii) coordinating and synthesizing horizontally across instances that involve discrepancies, inconsistencies, or redundancies. This proposal thus links up directly with modern agent-based computational models that stress the importance of so-called computational rationality in which the decision-making agent takes into account the computational costs involved in deciding to do things in one way versus another (e.g., Lieder & Griffiths, 2020). Re-representation is a way for the organism to reduce computational costs in future decision-making by reorganizing and re-representing elements in the knowledge base in more effective ways.

The need to eliminate discrepancies and redundancies in one's knowledge base (either within one's knowledge base or between it and current experience) is a major impetus to cognitive creativity and growth for both scientists and young children alike. To resolve the discrepancy between two conflicting perspectives—for example, I see you trying to retrieve your toy in A but I know it is in B—the thinking subject must create some new conceptualization, for example, the distinction between subjective beliefs and the objective situation. Indeed, it is through some such process that constructive thinking and re-representation help to solve the general problem of developmental discontinuity in Bayesian learning theories. Carey (2011, p. 120) formulates the problem and a potential solution as follows: "In cases of developmental discontinuity, the learner does not initially have the representational resources to state the hypotheses that will be tested, to represent the variables that could be associated or could be input to a Bayesian learning algorithm. Quinian bootstrapping [involving placeholders] is one learning process that can create new representational machinery, new concepts that articulate hypotheses previously unstatable." In the current agency-based model, it is the processes of constructive thinking and re-representation—coordinating and synthesizing existing conceptual material in novel ways (including via analogy)—that creates the new representational machinery and concepts enabling the articulation of previously unstatable hypotheses. Linguistic placeholders may play a prompting or guiding role in the process, but they are not strictly necessary; in all cases, the heavy lifting is done by creative processes of constructive thinking and re-representation.

Infants cannot engage in constructive thinking or re-representation because they are operating solely on the operational level of action and attention. With the emergence of an executive tier, toddlers are able to employ constructive thinking and executive re-representation to formulate novel theories and hypotheses. For example, in using tools toddlers observe the different ways that different tools work to solve particular problems, and through creative thinking may formulate novel hypotheses for novel tool-use situations. In addition, within a single problem, or perhaps across problems, toddlers detect, sometimes offline, inconsistencies or inefficiencies in their hypotheses about the causal principles that determine which tools do and do not work in which situations, and to resolve these they executively re-represent their local hypotheses to create a better overall theory of the causal structure of tool use more generally. They do this by coordinating conceptualizations—for example, tools of one type are better in one kind of situation and tools of another type are better in another kind of situation—and they generalize by abstracting and finding analogies. As another example, in learning a language, toddlers coordinate their existing linguistic symbols and constructions to create novel utterances and then executively re-represent these in more consistent and abstract representations by analogizing across word-based constructions to create more abstract linguistic representations.

Metacognitive re-representation operates one tier higher, as it were, as preschoolers are able to reflect on their thinking and planning, as well as their coordination of perspectives with others in shared agencies. Reflecting on and coordinating the operation of shared agencies leads to the construction of conceptual perspectives in which the same event or entity is considered as different from different conceptual perspectives simultaneously; for example, a toy may be simultaneously a dollhouse and a scale model, or an Arabic numeral simultaneously represents its cardinal and its ordinal values, or an object may be classified either by shape or by color. Metacognitive re-representation thus enables children to construct via horizontal coordinations and vertical integrations multi-perspectival concepts such as natural number or physical symbol. With the emergence of preschoolers' skills and motivations of collective intentionality, they are able to construct ideal universalized perspectives—what one ideally ought to believe or do—as a special type of perspective that can be integrated with others to yield objective representations such as beliefs or normative representations such as fairness. To construct these concepts requires both maturation of the child's agentive/cognitive architecture—including collective agency/intentionality—as well as particular kinds of experiences in their physical and sociocultural worlds. But, in addition, it also requires processes of constructive thinking and metacognitive re-representation in which children reflect on both their coordination of perspectives in shared agencies and the shared normative ideals of their collective cultural agencies.

Although there can be disagreements about specifics, I believe that almost all theorists would agree that children's executive and metacognitive monitoring and reflection on their own knowledge plays an important role in cognitive development. Using the analogy of child as scientist, it is clear that scientists often make conceptual progress by reflecting on their observations and thinking outside the laboratory. Nevertheless, these neglected processes of cognitive development—sometimes occurring offline outside of relevant decision-making and action—have been very little investigated. The current model involving two tiers of executive processes (executive and metacognitive) and two dimensions of re-representation (vertical abstractions and horizontal coordinations)—working not only in the formulation of hypotheses but also in the reorganization of existing knowledge—is an attempt to provide a starting point for more systematic empirical investigation.

8.3. Looking Ahead

Modern cognitive-developmental theorists often characterize their major question as: What is built-in and how do we learn the rest? (Spelke's recently published book is named *What Infants Know* and her planned second volume is tentatively titled *How Children Learn*.) This is fine as far as it goes, but I believe that what is built-in includes not just young infants' core concepts—as cognitive content—but also the agentive and cognitive architecture that is developing in significant ways during early ontogeny. All of the learning in which human children engage in the first years of life is structured by this architecture—its formats of cognitive representation and types of rational inference and self-regulation/reflection—both in terms of the kinds of experience that are possible and the kinds of learning processes that are available.

Children's Bayesian learning depends on wider competencies that change in significant ways over the first years of life.

8.3.1. Becoming Human

In my 2019 book, *Becoming Human*, I focused on those aspects of children's cognitive and social development that most clearly distinguish the human species from other great apes. I did this separately for eight different developmental pathways: social cognition, communication, cultural learning, cooperative thinking, collaboration, prosociality, social norms, and moral identity. In each case there appeared to be two developmental transitions: one at around nine months of age and another at around three years of age. I attributed the first transition to the emergence of skills and motivations of joint agency/intentionality (nine-month revolution) and the second transition to the emergence of skills and motivations of collective agency/intentionality (three-year normative turn). I reviewed much empirical evidence for these two transitions in each of the eight domains, often with systematic negative evidence for children below the age of the relevant transition.

What I have attempted to do here is to examine children's cognitive development more broadly—not just the species-unique sociocultural aspects, but the more individual dimensions of cognitive development as well, including those focused on the physical world. For this I borrowed from my 2022 book on the *Evolution of Agency,* which examines how humans' ancient animal ancestors evolved the cognitive skills necessary to engage in the agentive decision-making and action required for them to survive and thrive in their respective ecological niches. The result is the more general account of this book that takes a more comprehensive look at children's overall cognitive development, with its unique sociocultural dimension being only one aspect (and posited processes of "self-regulation" fleshed out in more detail under the rubrics of "constructive thinking" and "re-representation").

It turns out that the species-unique sociocultural dimension of human development requires supporting cognitive architectures at certain points, for example, imaginative representations to conceptualize joint intentional perspectives and metacognitive representations to conceptualize objective/normative perspectives. The discovery of this correspondence—or, if "discovery" is too strong a word, the hypothesizing of this correspondence—is what provided the unifying framework for the overall agentive account described in these pages.

8.3.2. Predictions and Questions

A theory is only as good as its predictions and the new questions it raises. As for predictions, Popper's (1959) well-known proposal is that the most powerful hypotheses are those that are most easily falsified. Almost all of the predictions I have made here are easily falsified, most directly through cross-cultural research or training studies that show children learning things before the current model says they are ready. Thus, although cross-cultural differences in children's cognitive development are to be expected, the prediction is that these fall within the constraints of the model. So if infants or toddlers could be trained to acquire skills or concepts beyond those specified in the model for their age—e.g., to engage prematurely in metacognitive or collaborative decision-making, or acquire universalized or normative concepts—this would provide strong evidence against the model. Of course, as always, there would need to be debate about such things as whether some demonstrated skill was "the real thing" or only some temporary or narrow version of the real thing, but science has a way of settling such issues over time.

More positively, the current agency-based model contributes to the making of predictions about the age at which children can acquire certain knowledge, concepts, and skills, albeit, admittedly, in only a general way. Thus, in the current model, the age at which a child will acquire some particular knowledge or concept is specified by the formula in Figure 8.4.

Agentive Architecture [capacities]	+	Specific Cognitive Prerequisites	+	Learning+ Re-representation	=>	Specific Knowledge & Concepts

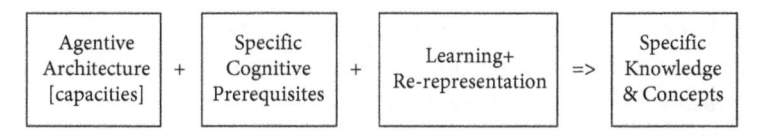

Figure 8.4. Summary of factors determining the age of emergence of any given knowledge or concept.

It is true that the agentive architecture is not very helpful in specifying particular ages; that is done at a much finer-grained level by specifying the prerequisite cognitive abilities and the particular learning experiences of the child. Nevertheless, the model specifies an important and theoretically significant parameter.

The current agency-based model also generates many new questions, a number of which have been asked in the preceding chapters. These have been mostly fairly specific questions, and so here I would like to lay out some of the larger sets of questions that the model raises for cognitive-developmental theory.

- Most cognitive-developmental theorists invoke evolutionary bases for at least some parts of their theory. It would therefore be very useful to know more about the ontogeny of agency and cognition in humans' evolutionary relatives at various degrees of remove (as represented by contemporary model species). In particular, very little is known about the cognitive development of infant and juvenile great apes and other nonhuman primates. And so the specific question is: how do the various developmental pathways in human cognitive development compare with those of humans' closest living relatives?
- Modern cognitive-developmental theories have not directly confronted the relation between methodology and theory. Specifically, different theories often rely on different methods to establish children's cognitive skills and knowledge, with researchers of infancy often using looking-time measures, researchers of toddlerhood often using problem-solving and behavioral choice measures, and researchers of early childhood often asking children to answer verbal questions. And so the general question is: is the knowledge that these different methods reveal of the same general nature or does it differ qualitatively?
- Modern cognitive-developmental theories all recognize the importance of both executive function and metacognition, but very few relate them to one another theoretically—as different types or manifestations of self-regulatory processes—or specify the complementary contributions they make to children's cognitive development at different ages. And so the general question is: how do executive function and metacognition—as two different types of self-regulatory processes—contribute to children's cognitive development at different ages?
- Modern cognitive-developmental theory traditionally has paid relatively little attention to processes of behavioral decision-making. Given the evolutionary importance of behavioral decision-making, this is a serious lacuna. The general questions here may be formulated in terms of the current agency-based model: Are infants confined to go/no-go decision-making? Are toddlers confined to go/no-go decision-making, as some theorists argue? Can toddlers engage in reflective decision-making and subsequent belief revision?
- Although modern theorists have made many attempts to specify the nature of infant cognitive representations, considerably less effort has been directed at the nature of toddler and preschooler cognitive representations. General questions here are: If toddlers are able to imaginatively represent non-actual situations, what are the limits of this ability

in terms of other types of counterfactual representations (e.g., can they represent impossible situations)? Are toddlers capable of constructing any types of multi-perspectival or objective/normative representations at all, and if not, why not?
- Despite much research on children's understanding of the launching event, there is very little research relating infants' and toddlers' understanding of the causal efficacy of their own actions and that of external causal forces. Does infants' sensitivity to the canonical nature of the launching event indicate an understanding of causal forces, and if so how does this relate to their understanding of the causal efficacy of their own actions on the world (and perhaps the causal efficacy of others' actions on the world)?
- Despite the centrality of processes of learning in all theories of cognitive development, relatively little attention has been paid to different types of learning. In particular, in the current model infants only attend to and learn fairly straightforward contingencies, whereas toddlers are more actively attempting to establish causal and intentional relations through active hypothesis testing, and preschoolers engage in metacognitive, pedagogical, and collaborative learning. Why are different types of learning associated with different ages, and what are their developmental sources?
- Some theorists have questioned whether young children's skills of shared intentionality actually begin with joint intentionality and joint attention and only after three years of age become collective and group-minded. Is infants' and toddlers' third-party sensitivity to group membership (as demonstrated by looking time and preference-based measures) the same thing as understanding oneself to be a member of a group and identifying with it? If not, what is the difference and where does it come from?
- Most theorists recognize the importance of processes of joint attention and shared intentionality in some aspects of children's lives, but the current emphasis on (recursive) mental coordination as the main cognitive operation necessary for creating and maintaining shared intentional interactions has been little studied. What are the different types of mental coordination, and do they in fact arise in shared intentional interactions, and is their nature fundamentally recursive?
- Although there is increasing interest in children's so-called learning by thinking, there has been relatively little empirical or theoretical work on the process—which I have called here constructive thinking and re-representation. We may thus ask not only about the nature of different

processes of these types, but also how they relate to one another and how they might change across age.
- The focus in the current account has been on species-universal aspects of human cognitive development. However, the cross-cultural research used to support inferences of universality is limited. We may therefore ask if there are cross-cultural variations—either in the existence or timing of the agentive architectures posited—that have been neglected in the account, and if so, what is their significance?

Many more specific questions could be raised, in particular with reference to the specifics of the major organizational transitions I have proposed at nine months and three years of age. In general, investigating the nature of these transitions and their different agentive and cognitive architectures should lead to important new insights into why and how children learn what they do when they do.

8.3.3. On to Middle Childhood

The first six years of life are in many ways the age of innocence. Across all human cultures, the end of early childhood at age six or seven is the age at which children are expected to begin mastering the tasks they will need to perform as adults in the culture. Indeed, in many cultures children of this age actually contribute to adult tasks in significant ways, as this is the age at which children have reached a level of competence enabling them to tend the sheep, gather firewood, take a message to the neighboring village, or help with household tasks (Kramer, 2005). In cultures that have formal education, it is the age at which literacy training formally begins and the age at which children are expected to be reasonable and responsible (if not yet fully competent) participants in a variety of community events. In many societies, age six or seven is the age at which children begin to be held responsible for their actions to some degree legally. As children transition to middle childhood the major life task is to become fully cultural agents, capable of pulling one's own weight responsibly in many different kinds of cultural activities.

Success in this task of course requires elaboration of the knowledge and cognitive skills developed during early childhood. Children in middle childhood become much more competent at combining their cognitive skills and using them to acquire novel skills. For example, their skills of reasoning in complex situations with multiple motives and/or perspectives involving counterfactual possibilities develop significantly in the early part of middle childhood (see Kushnir, 2022, for a review), enabling them to engage in reason-based deliberation in their decision-making. But my working hypothesis for the moment is that these older children are still working with the same three-tiered agentive architecture first constructed at age three; they are just increasing in their ability to coordinate more and more complicated things and to synthesize them in more and more complicated ways. One exceptional new skill may be the way that older children become capable of further applying their skills of perspective-taking and metacognitive reflection recursively—based on the possibility of individuals taking one another's perspectives on their perspectives indefinitely—so that they can reflect indefinitely on their own and others' thinking, and so construct ever broader and more "objective" perspectives on things.

9
The Child-as-Scientist Revisited

An agency-based model, informed by evolutionary analyses, holds the promise of bringing a new level of coherence to the study of human cognitive development. The particular agency-based model I have proposed here does this in a way that is fully compatible with modern cognitive-developmental theory; it simply emphasizes and organizes things in a somewhat different manner. In particular, the current agency-based model is fully compatible with modern characterizations of the child-as-scientist formulating theories, making Bayesian predictions, and actively testing those predictions through observation and experimentation. The child-as-scientist is an apt and very powerful metaphor, contributing to new and important discoveries on a regular basis. But apt and powerful does not mean fully complete.

As it exists in cognitive-developmental theory today, the child-as-scientist metaphor is incomplete in three basic ways. First, scientific progress is not uniformly continuous. Kuhn (1962) famously claimed that scientific progress takes place in revolutions in which new theories provide new lenses for viewing reality that are incommensurate with older ways of viewing things. It turns out that this is too radical a formulation, and there are both continuities and discontinuities as we move from, for example, Newtonian to Einsteinian physics or from Behaviorism to Cognitive Science. My claim here is not that the transitions from infancy to toddlerhood to early childhood represent incommensurate breaks, and it is not that everything happens all at once across the board instantaneously. My claim is simply that over early ontogeny new architectural organizations, heralding new cognitive capacities, emerge and are available to children in ways that enable them to cognitively represent, make inferences, and learn from their experience in new ways. Simply put, there are systematic differences among infants, toddlers, and preschool youngsters in the ways they understand and relate to the world. Changes in children's cognitive architectures, just as changes in scientific theories, are best described as qualitatively distinct steps with some continuities.

Second, as it is currently formulated, the child-as-scientist metaphor mostly addresses the use and testing of theories and hypotheses once they are formulated. But where do the theories and hypotheses come from in the first place? On one level, the sources of human creativity, including scientific creativity, are mysterious. But on another level, in the case of science one dimension of these sources is clear: scientists form theories by reflecting on and re-representing the scientific knowledge they already possess. They begin every hypothesis-testing article and every theoretical proposal with a review of the literature, which comprises attempts to abstract, coordinate, and synthesize existing knowledge. If existing scientific knowledge acts as Bayesian priors for experiments, the organization of these priors into theories is a reflective process, that is, a metacognitive process of the type characteristic of children from early childhood on. And metacognitive processes, as well as other executive processes, require an organizational architecture in which knowledge can be re-represented as needed to create novel concepts or to eliminate inconsistencies and inefficiencies in the knowledge base. Cognitive development is not just a process of observation and experimentation but also a process of reflection to create new conceptualizations. Both scientists and children develop cognitively both by using theories to observe and experiment on the world and also, in addition, by reflecting on their existing knowledge to create new theories moving forward.

Third, the child-as-scientist metaphor mostly ignores the fact that science is a social activity. This is true on two levels. On the one hand, modern science as a practice is a fundamentally collaborative activity in which the basic unit is the scientific team. A cursory glance at papers in major scientific journals suggests that the vast majority are published with multiple authors: the basic process is one of collaborative problem-solving and perspectival coordination with collaborators. Reflecting on her own scientific development, Carey (2022) notes that: "All of the progress I have ever made in understanding conceptual development has been hammered out in discussions, often heated arguments, with graduate student collaborators, postdoc collaborators, and faculty and research associate collaborators." In addition, on another level, it has been clear at least since Kuhn (1962) and Lakatos (1970) that modern science is a cultural institution, or set of institutions, in the sense that individual scientists work in socially constructed theoretical paradigms that set questions, define key concepts, and stipulate methods and standards of evidence as social norms of scientific inquiry (e.g., in textbooks). Junior scientists enter the field mostly by serving as apprentices in the laboratory of senior scientists, who enculturate them into the ways of the paradigm. Revolutions in scientific paradigms are often effected by individuals, but these are individuals who are embedded in the concepts and procedures that are collectively accepted and transmitted in their field. The claim in this case is that junior scientists and children develop cognitively both by interacting jointly with collaborative partners and by interacting collectively within normative institutional conventions—typically as introduced to them through the practices of mature scientists/adults—which together provide the sociocultural context within which they progress.

If the child is a kind of scientist—as the modern theory proposes—then I am only suggesting that we explicitly acknowledge that (1) scientific change sometimes occurs in qualitative shifts; (2) progress in science depends not only on the bedrock of observation and experimentation but also on periodic theoretical reorganizations (i.e., re-representations) of already existing observations and concepts; and (3) scientific progress is due not only to individual agency and cognition, but also to the individual's mental coordination with other scientists in the context of the shared questions, concepts, and methods of a collectively constituted scientific paradigm. The hope is that by broadening the modern theory of human cognitive development in these ways, we will make further progress in understanding how children as scientists come to learn about and make sense of their worlds.

References

Akhtar, N. (1999). Acquiring basic word order: Evidence for data-driven learning of syntactic structure. *Journal of Child Language, 26,* 339–356.
Akhtar, N., & Tomasello, M. (1996). 2-year-olds learn words for absent objects and actions. *British Journal of Developmental Psychology, 14,* 79–93.
Alderete, S., & Xu, F. (2023). Three-year-old children's reasoning about possibilities. *Cognition, 237,* 105472.
Baillargeon, R. (1987). Object permanence in 3½-and 4½-month-old infants. *Developmental Psychology, 23*(5), 655.
Baillargeon, R. (2008). Innate ideas revisited: For a principle of persistence in infants' physical reasoning. *Perspectives on Psychological Science, 3*(1), 2–13.
Bard, K. (2012). Emotional engagement: How chimpanzee minds develop. In: F. De Waal & P. Ferrari, (Eds.), *The Primate Mind: Built to Engage with Other Minds* (pp. 224–245). Cambridge, MA: Harvard University Press.
Barsalou, L. W. (2008). Grounded cognition. *Annual Review Psychology, 59,* 617–645.
Bartsch, K., & Wellman, H. (1995). *Children Talk About the Mind.* New York: Oxford University Press.
Bates, E. (1979). *The Emergence of Symbols.* New York: Academic Press.
Bauer, P., Schwade, J., Wewerka, S., & Delaney, K. (1999). Planning ahead: Goal-directed problem solving by 2-year-olds. *Developmental Psychology, 35*(5), 1321.
Behne, T., Carpenter, M., & Tomasello, M. (2005). 1-year-olds comprehend the communicative intentions behind gestures in a hiding game. *Developmental Science, 8,* 492–499.
Behne, T., Liszkowski, U., Carpenter, M., & Tomasello, M. (2012). Twelve-month-olds' comprehension and production of pointing. *British Journal of Developmental Psychology, 30,* 359–375.
Bellagamba, F., & Tomasello, M. (1999). Re-enacting intended acts: Comparing 12-and 18-month-olds. *Infant Behavior and Development, 22*(2), 277–282.
Beran, M., Perner, J., & Proust, J. (Eds.). (2012). *Foundations of Metacognition.* Oxford: Oxford University Press.
Bergelson, E., & Swingley, D. (2012). At 6–9 months, human infants know the meanings of many common nouns. *Proceedings of the National Academy of Sciences, 109*(9), 3253–3258.
Berkman, E., Hutcherson, C., Livingston, J., Kahn, L., & Inzlicht, M. (2017). Self-control as value-based choice. *Current Directions in Psychological Science, 26,* 422–428.
Bermudez, J. (2003). *Thinking without Words.* Oxford University Press.
Blakey, E., Visser, I., & Carroll, D. (2016). Different executive functions support different kinds of cognitive flexibility: Evidence from 2-, 3-, and 4-year-olds. *Child Development, 87,* 513–526.
Bloom, L., & Capatides, J. (1987). Sources of meaning in the acquisition of complex syntax: The sample case of causality. *Journal of Experimental Child Psychology, 43*(1), 112–128.
Bloom, P. (2002). *How Children Learn the Meanings of Words.* Cambridge, MA: MIT Press.

Bohn, M., Call, J., & Tomasello, M. (2019). Natural reference: A phylo-and ontogenetic perspective on the comprehension of iconic gestures and vocalizations. *Developmental Science, 22*(2), e12757.

Bohn, M., Kordt, C., Braun, M., Call, J., & Tomasello, M. (2020). Learning novel skills from iconic gestures: A developmental and evolutionary perspective. *Psychological Science, 31*(7), 873–880.

Bonawitz, E., Ferranti, D., Saxe, R., Gopnik, A., Meltzoff, A., Woodward, J., & Schulz, L. E. (2010). Just do it? Investigating the gap between prediction and action in toddlers' causal inferences. *Cognition, 115*(1), 104–117.

Bonner, J. (1988). *The Evolution of Complexity by Means of Natural Selection*. Princeton University Press.

Bratman, M. (2014). *Shared Agency: A Planning Theory of Acting Together*. New York: Oxford University Press.

Brownell, C., & Carriger, M. (1990). Changes in cooperation and self-other differentiation during the second tear. *Child Development, 61*, 1164–1174.

Bulley, A., McCarthy, T., Gilbert, S. J., Suddendorf, T., & Redshaw, J. (2020). Children devise and selectively use tools to offload cognition. *Current Biology, 30*(17), 3457–3464.

Burge, T. (2011). Disjunctivism again. *Philosophical Explorations, 14*(1), 43–80.

Butler, L. (2020). The empirical child? A framework for investigating the development of scientific habits of mind. *Child Development Perspectives, 14*(1), 34–40.

Butler, L., & Tomasello, M. (2016). Two-and 3-year-olds integrate linguistic and pedagogical cues in guiding inductive generalization and exploration. *Journal of Experimental Child Psychology, 145*, 64–78.

Buttelmann, D., Carpenter, M., Call, J., & Tomasello, M. (2007). Enculturated chimpanzees imitate rationally. *Developmental Science, 10*(4), F31–F38.

Buttelmann, D., Zmyj, N., Daum, M., & Carpenter, M. (2013). Selective imitation of in-group over out-group members in 14-month-old infants. *Child Development, 84*, 422–428.

Call, J. (2000). Estimating and operating on discrete quantities in orangutans. *Journal of Comparative Psychology, 114*(2), 136.

Call, J. (2004). Inferences about the location of food in the great apes (*Pan paniscus*, *Pan troglodytes*, *Gorilla gorilla*, and *Pongo pygmaeus*). *Journal of Comparative Psychology, 118*(2), 232–241.

Call, J., & Carpenter, M. (2001). Do apes and children know what they have seen? *Animal Cognition, 3*(4), 207–220.

Call, J., Hare, B., Carpenter, M., & Tomasello, M. (2004). "Unwilling" versus "unable": Chimpanzees' understanding of human intentional action. *Developmental Science, 7*(4), 488–498.

Call, J., & Tomasello, M. (in press). *Primate cognition, 2nd Edition*. Oxford Univeristy Press.

Call, J., & Tomasello, M. (2008). Does the chimpanzee have a theory of mind? 30 years later. *Trends in Cognitive Sciences, 12*(5), 187–192.

Callaghan, T., Moll, H., Rakoczy, H., Warneken, F, Liszkowski, U. Behne, T., & Tomasello, M. (2011). Early social cognition in three cultural contexts. *Monographs of the Society for Research in Child Development, 76*(2), 1–142.

Campos, J., Anderson, D., Barbu-Roth, M., Hubbard, E., Hertenstein, M., & Witherington, D. (2000). Travel broadens the mind. *Infancy, 1*(2), 149–219.

Carey, S. (1985). *Conceptual Change in Childhood*. Cambridge, MA: MIT Press.

Carey, S. (2009). *The Origin of Concepts*. Oxford University Press.

Carey, S. (2011). Précis of the origin of concepts. *Behavioral and Brain Sciences, 34*(3), 113–124.

Carey, S. (2022). Becoming a cognitive scientist. *Annual Review of Developmental Psychology*, 4, 1–19.

Carlson, S. M. (2023). Let me choose: The role of choice in the development of executive function skills. *Current Directions in Psychological Science*, 32(3), 220–227.

Carpenter, M., Call, J., & Tomasello, M. (2002). Some 36-month-old children understand false beliefs. *British Journal of Developmental Psychology*, 20, 393–420.

Carpenter, M., Nagell, K., Tomasello, M., Butterworth, G., & Moore, C. (1998). Social cognition, joint attention, and communicative competence from 9 to 15 months of age. *Monographs of the Society for Research in Child Development*, i–174.

Charman, T., Baron-Cohen, S., Swettenham, J., Baird, G., Cox, A., & Drew, A. (2000). Testing joint attention, imitation, and play as infancy precursors to language and theory of mind. *Cognitive Development*, 15(4), 481–498.

Cheung, P., Toomey, M., Jiang, Y., Stoop, T., & Shusterman, A. (2022). Acquisition of the counting principles during the subset-knower stages: Insights from children's errors. *Developmental Science*, 25(4), e13219.

Clark, H. (1996). *Using Language*. Cambridge University Press.

Crane, T. (2003). *The Mechanical Mind: A Philosophical Introduction to Minds, Machines and Mental Representation* (2nd ed.). New York: Routledge.

Csibra, G., & Gergely, G. (2009). Natural pedagogy. *Trends in Cognitive Sciences*, 13(4), 148–153.

Darwall, S. (2006). *The Second-Person Standpoint: Respect, Morality, and Accountability*. Cambridge, MA: Harvard University Press.

De Villiers, J., & Pyers, J. (2002). Complements to cognition: A longitudinal study of the relationship between complex syntax and false-belief-understanding. *Cognitive Development*, 17, 1037–1060.

DeLoache, J. S. (2004). Becoming symbol-minded. *Trends in Cognitive Sciences*, 8(2), 66–70.

Devine, R., & Hughes, C. (2014). Relations between false belief understanding and executive function in early childhood: A meta-analysis. *Child Development*, 85(5), 1777–1794.

Diamond, A. (1985). Development of the ability to use recall to guide action, as indicated by infants' performance on AB. *Child Development*, 56(4), 868–883.

Diamond, A. (1990). Developmental time course in human infants and infant monkeys, and the neural bases of, inhibitory control in reaching. *Annals of the New York Academy of Sciences*, 608(1), 637–676.

Diamond, A. (2013). Executive functions. *Annual Review of Psychology*, 64, 135–168.

Diamond, A., & Gilbert, J. (1989). Development as progressive inhibitory control of action: Retrieval of a contiguous object. *Cognitive Development*, 4(3), 223–249.

Diaz, V., & Farrar, M. (2017). The missing explanation of bilinguals false-belief advantage: A longitudinal study. *Developmental Science*, 21(4), e12594.

Dickinson, A. (2001). Causal learning: An associative analysis (The 28th Bartlett Memorial Lecture). *Quarterly Journal of Experimental Psychology*, 54B, 3–25.

Diesendruck, G. (2005). The principles of conventionality and contrast in word learning: An empirical examination. *Developmental Psychology*, 41(3), 451.

Doebel, S. (2020). Rethinking executive function and its development. *Perspectives on Psychological Science*, 15(4), 942–956.

Doebel, S., & Zelazo, P. (2015). A meta-analysis of the Dimensional Change Card Sort: Implications for developmental theories and the measurement of executive function in children. *Developmental Review*, 38, 241–268.

Doherty, M., & Perner, J. (1998). Metalinguistic awareness and theory of mind: Just two words for the same thing? *Cognitive Development*, 13, 279–305.

Duguid, S., Wyman, E., Bullinger, A., Herfurth, K., & Tomasello, M. (2014). Coordination strategies of chimpanzees and human children in a Stag Hunt game. *Proceedings of the Royal Society B: Biological Sciences, 281*(1796), 20141973.

Duguid, S., Wyman, E., Grüneisen, S., & Tomasello, M. (2020). The strategies used by chimpanzees (*Pan troglodytes*) and children (*Homo sapiens*) to solve a simple coordination problem. *Journal of Comparative Psychology, 134*(4), 401.

Dunham, Y. (2018). Mere membership. *Trends in Cognitive Sciences, 22*(9), 780–793.

Egner, T. (2017). *The Wiley Handbook of Cognitive Control*. Chichester: Wiley-Blackwell.

Elsner, B. (2007). Infants' imitation of goal-directed actions: The role of movements and action effects. *Acta Psychologica, 124*(1), 44–59.

Engelmann, J. M., & Tomasello, M. (2019). Children's sense of fairness as equal respect. *Trends in Cognitive Sciences, 23*(6), 454–463.

Engelmann, J. M., Völter, C. J., Goddu, M. K., Call, J., Rakoczy, H., & Herrmann, E. (2023). Chimpanzees prepare for alternative possible outcomes. *Biology Letters, 19*(6), 20230179.

Engelmann, J. M., Völter, C., O'Madagain, C., Proft, M., Haun, D., Rakoczy, H., & Herrmann, E. (2021). Chimpanzees consider alternative possibilities. *Current Biology, 31*(20), R1377–R1378.

Fizke, E., Barthel, D., Peters, T., & Rakoczy, H. (2014). Executive function plays a role in coordinating different perspectives, particularly when one's own perspective is involved. *Cognition, 130*(3), 315–334.

Flavell, J. (1992). Perspectives on perspective taking. In: H. Beilin & P. B. Pufall (Eds.), *Piaget's Theory: Prospects and Possibilities* (pp. 107–139). Mahwah: Lawrence Erlbaum Associates, Inc.

Flavell, J., Everett, B., Croft, K., & Flavell, E. (1981). Young children's knowledge about visual perception: Further evidence for the Level 1–Level 2 distinction. *Developmental Psychology, 17*(1), 99–103.

Frank, M., Everett, D., Fedorenko, E., & Gibson, E. (2008). Number as a cognitive technology: Evidence from Pirahã language and cognition. *Cognition, 108*(3), 819–824.

Gelman, R., & Gallistel, C. (1978). *The Child's Understanding of Number*. Cambridge, MA: Harvard University Press.

Gelman, S. A., & Bloom, P. (2007). Developmental changes in the understanding of generics. *Cognition, 105*, 166–183.

Gentner, D. (2003). Why we're so smart. In: D. Gentner & S. Goldin-Meadow (Eds.), *Language in Mind: Advances in the Study of Language and Thought* (pp. 195–235). Cambridge, MA: MIT Press.

Gentner, D. (2010). Bootstrapping the mind: Analogical processes and symbol systems. *Cognitive Science, 34*(5), 752–775.

Geraci, A., & Surian, L. (2011). The developmental roots of fairness: Infants' reactions to equal and unequal distributions of resources. *Developmental Science, 14*, 1012–1020.

Gergely, G., Bekkering, H., & Király, I. (2002). Rational imitation in preverbal infants. *Nature, 415*, 755.

Gershman, S., Horvitz, E., & Tenenbaum, J. (2015). Computational rationality: A converging paradigm for intelligence in brains, minds, and machines. *Science, 349*, 273–278.

Gerson, S., & Woodward, A. (2014). Learning from their own actions: The unique effect of producing actions on infants' action understanding. *Child Development, 85*(1), 264–277.

Goddu, M., & Gopnik, A. (2024). The development of human causal learning and reasoning. *Nature Reviews Psychology*, 2024, 1–21.

Goldberg, A. (2003). Constructions: A new theoretical approach to language. *Trends in Cognitive Sciences, 7*(5), 219–224.

Goldin-Meadow, S. (2005). *Hearing Gesture: How Our Hands Help Us Think*. Cambridge, MA: Harvard University Press.
Gómez, J. C. (2006). *Apes, Monkeys, Children, and the Growth of Mind*. Cambridge, MA: Harvard University Press.
Gopnik, A. (2020). Childhood as a solution to explore–exploit tensions. *Philosophical Transactions of the Royal Society B*, *375*(1803), 20190502.
Gopnik, A., & Meltzoff, A. (1997). *Words, Thoughts, and Theories*. Cambridge, MA: MIT Press.
Gopnik, A., Sobel, D., Schulz, L., & Glymour, C. (2001). Causal learning mechanisms in very young children: Two-, three-, and four-year-olds infer causal relations from patterns of variation and covariation. *Developmental Psychology*, *37*(5), 620–629.
Gopnik, A., & Wellman, H. (2012). Reconstructing constructivism: Causal models, Bayesian learning mechanisms, and the theory theory. *Psychological Bulletin*, *138*, 1085–1108.
Gottlieb, G. (2007). Probabilistic epigenesis. *Developmental Science*, *10*(1), 1–11.
Gould, S. (1977). *Ontogeny and Phylogeny*. Cambridge, MA: Belknap Press of Harvard University Press.
Goupil, L., & Proust, J. (2023). Curiosity as a metacognitive feeling. *Cognition*, *231*, 105325.
Goupil, L., Romand-Monnier, M., & Kouider, S. (2016). Infants ask for help when they know they don't know. *Proceedings of the National Academy of Sciences*, *113*(13), 3492–3496.
Gräfenhain, M., Behne, T., Carpenter, M., & Tomasello, M. (2009). Young children's understanding of joint commitments. *Developmental Psychology*, *45*(5), 1430–1443.
Gräfenhain, M., Carpenter, M., & Tomasello, M. (2013). 3-year-olds' understanding of the consequences of joint commitments. *PLoS One*, *8*(9), e73039.
Grocke, P., Rossano, F., & Tomasello, M. (2015). Procedural justice in children: Preschoolers accept unequal resource distributions if the procedure provides equal opportunities. *Journal of Experimental Child Psychology*, *140*, 197–210.
Grosse, G., Carpenter, M., Tomasello, M., & Behne, T. (2010). Infants Communicate in Order to Be Understood. *Developmental Psychology*, *46*(6), 1710–1722.
Grosse-Wiesmann, C., Friederici, A. D., Disla, D., Steinbeis, N., & Singer, T. (2017). Longitudinal evidence for 4-year-olds' but not 2- and 3-year-olds' false belief-related action anticipation. *Cognitive Development*, *20*, e12445.
Grüneisen, S., Wyman, E., & Tomasello, M. (2015a). Children use salience to solve coordination problems. *Developmental Science*, *18*, 495–501.
Grüneisen, S., Wyman, E., & Tomasello, M. (2015b). Conforming to coordinate: Children use majority information for peer coordination. *British Journal of Developmental Psychology*, *33*, 136–147.
Gweon, H. (2021). Inferential social learning: Cognitive foundations of human social learning and teaching. *Trends in Cognitive Sciences*, *25*(10), 896–910.
Hamann, K., Bender, J., & Tomasello, M. (2014). Meritocratic sharing is based on collaboration in 3-year-olds. Developmental Psychology, *50*(1), 121–128.
Hamann, K., Warneken, F., Greenberg, J. R., & Tomasello, M. (2011). Collaboration encourages equal sharing in children but not in chimpanzees. *Nature*, *476*(7360), 328–331.
Hamlin, K., Wynn, K., & Bloom, P. (2007). Social evaluation by preverbal infants. *Nature*, *450*(7169), 557–559.
Hanus, D., & Call, J. (2008). Chimpanzees infer the location of a reward on the basis of the effect of its weight. *Current Biology*, *18*(9), R370–R372.

Hanus, D., & Call, J. (2011). Chimpanzee problem-solving: Contrasting the use of causal and arbitrary cues. *Animal Cognition, 14*(6), 871–878.

Hardecker, S., Schmidt, M., & Tomasello, M. (2017). Children's developing understanding of the conventionality of rules. *Journal of Cognition and Development, 18*(2), 163–188.

Hare, B., Call, J., Agnetta, B., & Tomasello, M. (2000). Chimpanzees know what conspecifics do and do not see. *Animal Behaviour, 59*, 771–785.

Hare, B., Call, J., & Tomasello, M. (2001). Do chimpanzees know what conspecifics know?. *Animal behaviour, 61*(1), 139–151.

Harris, P. L. (2000). *The work of the imagination.* Blackwell Publishing.

Hauf, P., Elsner, B., & Aschersleben, G. (2004). The role of action effects in infants' action control. *Psychological Research, 68*, 115–125.

Haun, D., & Tomasello, M. (2014). Children conform to the behavior of peers; Great apes stick with what they know. *Psychological Science, 25*, 2160–2167.

Herrmann, E., Misch, A., & Tomasello, M. (2015). Uniquely human self-control begins at school age. *Developmental Science, 18*, 979–993.

Hirschfeld, L., & Gelman, S. (Eds.). (1994). *Mapping the mind: Domain specificity in cognition and culture.* Cambridge University Press.

Hrdy, S. (2006). Evolutionary context of human development: The cooperative breeding model. In: Dahlem Workshop No. 92., C. Carter & L. Ahnert (Eds), *Attachment and Bonding: A New Synthesis* (pp. 19–46). Cambridge, MA: MIT Press.

Hrdy, S. (2016). Development plus social selection in the emergence of "emotionally modern" humans. In: C. L. Meehan & A. N. Crittenden (Eds.), *Childhood: Origins, Evolution, and Implications* (pp. 11–44). Albuquerque, NM: University of New Mexico Press.

Ibbotson, P. (2014). Little dictators: A developmental meta-analysis of prosocial behavior. *Current Anthropology, 55*(6), 814–821.

Inhelder, B., & Piaget, J. (1964). *The Early Growth of Logic in the Child: Classification and Seriation.* (E. A. Lunzer & D. Papert, Trans.). New York: Harper & Row.

Kachel, U., Svetlova, M., & Tomasello, M. (2018). 3-year-olds' reactions to a partner's failure to perform her role in a joint commitment. *Child Development, 89*, 1691–1703.

Kachel, U., & Tomasello, M. (2019). 3- and 5-year-old children's adherence to explicit and implicit joint commitments. *Developmental Psychology, 55*, 80–88.

Kalnins, I., & Bruner, J. (1973). The coordination of visual observation and instrumental behavior in early infancy. *Perception, 2*(3), 307–314.

Kaminski, J., Call, J., & Tomasello, M. (2004). Body orientation and face orientation: Two factors controlling apes' begging behavior from humans. *Animal Cognition, 7*(4), 216–223.

Kanngiesser, P., Mammen, M., & Tomasello, M. (2021). Young children's understanding of justifications for breaking a promise. *Cognitive Development, 60*, 101127.

Kano, F., & Call, J. (2014). Great apes generate goal-based action predictions: An eye-tracking study. *Psychological Science, 25*(9), 1691–1698.

Kant, I. (1781). *Critique of pure reason* (J. M. D. Meiklejohn, Trans.). Willey Book Co.

Karg, K., Schmelz, M., Call, J., & Tomasello, M. (2014). All great ape species and two-and-a-half-year-old children discriminate appearance from reality. *Journal of Comparative Psychology, 128*(4), 431–439.

Karg, K., Schmelz, M., Call, J., & Tomasello, M. (2015). The goggles experiment: Can chimpanzees use self-experience to infer what a competitor can see? *Animal Behavior, 105*, 211–221.

Karg, K., Schmelz, M., Call, J., & Tomasello, M. (2016). Differing views: Can chimpanzees do Level 2 perspective-taking? *Animal Cognition, 19*, 555–564.

Karmiloff-Smith, A. (1992). *Beyond Modularity: A Developmental Perspective on Cognitive Science*. Cambridge, Mass: MIT Press.

Kelemen, D. (2004). Are children "intuitive theists"? Reasoning about purpose and design in nature. *Psychological Science, 15*(5), 295–301.

Kim, S., Sodian, B., & Proust, J. (2020). 12-and 24-month-old infants' search behavior under informational uncertainty. *Frontiers in Psychology, 11*, 566.

Kimura, K., & Gopnik, A. (2019). Rational higher-order belief revision in young children. *Child Development, 90*(1), 91–97.

Koechlin, E., & Summerfield, C. (2007). An information theoretical approach to prefrontal executive function. *Trends in Cognitive Sciences, 11*, 229–235.

Köymen, B., O'Madagain, C., Domberg, A., & Tomasello, M. (2020). Young children's ability to produce valid and relevant counter-arguments. *Child Development, 91*(3), 685–693.

Köymen, B., Rosenbaum, L., & Tomasello, M. (2014). Reasoning during joint decision-making by preschool peers. *Cognitive Development, 32*, 74–85.

Köymen, B., & Tomasello, M. (2018). Children's meta-talk in their collaborative decision-making with peers. *Journal of Experimental Child Psychology, 166*, 549–566.

Köymen, B., & Tomasello, M. (2020). The early ontogeny of reason giving. *Child Development Perspectives, 14*(4), 215–220.

Krachun, C., Carpenter, M., Call, J., & Tomasello, M. (2009). A competitive nonverbal false belief task for children and apes. *Developmental Science, 12*(4), 521–535.

Kramer, K. (2005). Children's help and the pace of reproduction: Cooperative breeding in humans. *Evolutionary Anthropology: Issues, News, and Reviews: Issues, News, and Reviews, 14*(6), 224–237.

Kruger, A. (1992). The effect of peer and adult–child transactive discussions on moral reasoning. *Merrill-Palmer Quarterly, 38*(2), 191–211.

Krupenye, C., Kano, F., Call, J., Hirata, S., & Tomasello, M. (2016). Great apes anticipate that other individuals will act according to false beliefs. *Science, 354*, 110–114.

Kuhn, D., & Dean, D. (2004). A bridge between cognitive psychology and educational practice. *Theory into Practice, 43*(4), 268–273.

Kuhn, T. (1962). *The Structure of Scientific Revolutions*. Chicago, IL: University of Chicago Press.

Kushnir, T. (2022). Imagination and social cognition in childhood. *Wiley Interdisciplinary Reviews: Cognitive Science, 13*(4), e1603.

Lakatos, I. (1970). Falsification and the methodology of scientific research programmes. In: I. Lakatos, & A. Musgrave (Eds.), *Criticism and the Growth of Knowledge* (pp. 91–196). Cambridge, MA: Cambridge University Press.

Langer, J., & Killen, M. (1998). Phylogenetic and ontogenetic origins of cognition: Classification. In: J. Langer, & M. Killen (Eds.), *Piaget, Evolution, and Development* (pp. 41–62). New York: Psychology Press.

Leahy, B. (2023). Don't you see the possibilities? Young preschoolers may lack possibility concepts. *Developmental Science, 26*(6), e13400.

Leahy, B., & Carey, S. (2020). The acquisition of modal concepts. *Trends in Cognitive Sciences, 24*(1), 65–78.

Leavens, D., & Hopkins, W. (1998). Intentional communication by chimpanzees: A cross-sectional study of the use of referential gestures. *Developmental Psychology, 34*(5), 813.

Leslie, A. (1987). Pretense and representation: The origins of "theory of mind." *Psychological Review 94*, 412–426.

Lewis, D. (1969). *Convention*. Cambridge, MA: Harvard University Press.

Li, L., & Tomasello, M. (2022). Disagreement, justification, and equitable moral judgments: A brief training study. *Journal of Experimental Child Psychology, 223*, 105494.

Liebal, K., Behne, T., Carpenter, M., & Tomasello, M. (2009). Infants use shared experience to interpret pointing gestures. *Developmental Science, 12*(2), 264–271.

Liebal, K., Carpenter, M., & Tomasello, M. (2013). Young children's understanding of cultural common ground. *British Journal of Developmental Psychology, 31*(1), 88–96.

Lieder, F., & Griffiths, T. (2020). Resource-rational analysis: Understanding human cognition as the optimal use of limited computational resources. *Behavioral and Brain Sciences, 43*, e1.

Lieven, E., Behrens, H., Speares, J., & Tomasello, M. (2003). Early syntactic creativity: A usage-based approach. *Journal of Child Language, 30*(2), 333–370.

Liszkowski, U., Brown, P., Callaghan, T., Takada, A. and de Vos, C. (2012). A prelinguistic gestural universal of human communication. *Cognitive Science, 36*, 698–713.

Liszkowski, U., Carpenter, M., & Tomasello, M. (2007). Pointing out new news, old news, and absent referents at 12 months of age. *Developmental Science, 10*(2), F1–F7.

Liszkowski, U., Carpenter, M., & Tomasello, M. (2008). Twelve-month-olds communicate helpfully and appropriately for knowledgeable and ignorant partners. *Cognition, 108*(3), 732–739.

Lock, A. (1978). On being picked up. In: A. Lock (Ed.), *Action, Gesture and Symbol: The Emergence of Language*. London: Academic Press.

Lohmann, H., & Tomasello, M. (2003). The role of language in the development of false belief understanding: A training study. *Child Development, 74*(4), 1130–1144.

MacLean, E., Hare, B., Nunn, C., Addessi, E., Amici, F., Anderson, R., ... & Zhao, Y. (2014). The evolution of self-control. *Proceedings of the National Academy of Sciences, 111*(20), E2140–E2148.

Mandler, J. M. (1992). How to build a baby: II. Conceptual primitives. *Psychological Review, 99*(4), 587.

Mandler, J. (2004). *The Foundations of Mind: Origins of Conceptual Thought*. Oxford University Press.

Mandler, J. (2007). On the origins of the conceptual system. *American Psychologist, 62*(8), 741.

Marcovitch, S., & Zelazo, P. (2006). The influence of number of A trials on 2-year-olds' behavior in two A-not-B-type search tasks: A test of the hierarchical competing systems model. *Journal of Cognition and Development, 7*(4), 477–501.

Markman, A., & Stilwell, C. (2001). Role-governed categories. *Journal of Experimental & Theoretical Artificial Intelligence, 13*(4), 329–358.

Markman, E. (1989). *Categorization and Naming in Children*. Cambridge, MA: MIT Press.

Matthews, D., Behne, T., Lieven, E., & Tomasello, M. (2012). Origins of the human pointing gesture: A training study. *Developmental Science, 15*(6), 817–829.

Meltzoff, A. (1995). Understanding the intentions of others: Re-enactment of intended acts by 18-month-old children. *Developmental Psychology, 31*, 1–16.

Meltzoff, A. (2005). Imitation and other minds: The "like me" hypothesis. *Perspectives on Imitation: From Neuroscience to Social Science, 2*, 55–77.

Meltzoff, A. (2007). The "like me" framework for recognizing and becoming an intentional agent. *Acta Psychologica, 124*, 26–43.

Meltzoff, A., & Brooks, R. (2008). Self-experience as a mechanism for learning about others: A training study in social cognition. *Developmental Psychology, 44*(5), 1257.

Meltzoff, A., Waismeyer, A., & Gopnik, A. (2012). Learning about causes from people: Observational causal learning in 24-month-old infants. *Developmental Psychology, 48*(5), 1215.

Mendes, N., Rakoczy, H., & Call, J. (2008). Ape metaphysics: Object individuation without language. *Cognition, 106*(2), 730–749.

Millar, W., & Watson, J. (1979). The effect of delayed feedback on infant learning reexamined. *Child Development, 50*, 747–751.

Miller, G., Galanter, E., & Pribram, K. (1960). *Plans and the Structure of Behavior.* New York: Holt & Co.

Milligan, K., Astington, J., & Dack, L. (2007). Language and theory of mind: Metaanalysis of the relation between language ability and false-belief understanding. *Child Development, 78*(2), 622–646.

Mody, S., & Carey, S. (2016). The emergence of reasoning by the disjunctive syllogism in early childhood. *Cognition, 154*, 40–48.

Moll, H., & Meltzoff, A. (2011). How does it look? Level 2 perspective-taking at 36 months of age. *Child Development, 82*(2), 661–673.

Moll, H., Meltzoff, A., Merzsch, K., & Tomasello, M. (2013). Taking versus confronting visual perspectives in preschool children. *Developmental Psychology, 49*(4). 646–654.

Moll, H., & Tomasello, M. (2004). 12- and 18-month-old infants follow gaze to spaces behind barrier. *Developmental Science, 7*(1), F1–F9.

Moll, H., & Tomasello, M. (2006). Level 1 perspective-taking at 24 months of age. *British Journal of Developmental Psychology, 24*(3), 603–613.

Moll, H., & Tomasello, M. (2012). 3-year-olds understand appearance and reality—just not about the same object at the same time. *Developmental Psychology, 48*(4), 1124–1132.

Muentener, P., & Schulz, L. (2014). Toddlers infer unobserved causes for spontaneous events. *Frontiers in Psychology, 5*, 1496.

Murray, L., & Trevarthen, C. (1985). Emotional regulation of interactions between two-month- olds and their mothers. In: T. M. Field & N. A. Fox (Eds.), *Social Perception in Infants* (pp. 177–197). Norwood, NJ: Ablex Publishers.

Myowa-Yamakoshi, M., Tomonaga, M., Tanaka, M., & Matsuzawa, T. (2004). Imitation in neonatal chimpanzees (*Pan troglodytes*). *Developmental Science, 7*(4), 437–442.

Ng, R., Heyman, G., & Barner, D. (2011). Collaboration promotes proportional reasoning about resource distribution in young children. *Developmental Psychology, 47*(5), 1230.

Noyes, A., Keil, F., & Dunham, Y. (2020). Institutional actors: Children's emerging beliefs about the causal structure of social roles. *Developmental Psychology, 56*(1), 70.

Noyes, A., Keil, F., Dunham, Y. (2018). The emerging causal understanding of institutional objects. *Cognition, 170*, 83–87.

Nyhout, A., & Ganea, P. (2019). Mature counterfactual reasoning in 4-and 5-year-olds. *Cognition, 183*, 57–66.

O'Madagain, C., Helming, K., Schmidt, M., Shupe, E., Call, J., & Tomasello, M. (2022). Great apes and human children rationally monitor their decisions. *Proceedings of the Royal Society B, 289*(1971), 20212686.

O'Madagain, C., & Tomasello, M. (2021). Joint attention to mental content and the social origin of reasoning. *Synthese, 198*, 4057–4078.

Onishi, K., & Baillargeon, R. (2005). Do 15-month-old infants understand false beliefs? *Science, 308*, 255–258.

Perner, J., Brandl, J., & Garnham, A. (2003). What is a perspective problem? Developmental issues in understanding belief and dual identity. *Facta Philosophica, 5*, 355–378.

Perner, J., & Lang, B. (2002). What causes 3-year-olds' difficulty on the Dimensional Change Card Sorting task? *Infant and Child Development*, 11, 93–105.
Piaget, J. (1952). *The Origins of Intelligence in Children*. New York: Norton.
Piaget, J. (1954). *The Construction of Reality in the Child*. New York: Norton.
Podjarny, G., Kamawar, D., & Andrews, K. (2017). The Multidimensional Card Selection Task: A new way to measure concurrent cognitive flexibility in preschoolers. *Journal of Experimental Child Psychology*, 159, 199–218.
Podjarny, G., Kamawar, D., & Andrews, K. (2022). Two birds in the hand: Concurrent and switching cognitive flexibility in preschoolers. *Journal of Experimental Child Psychology*, 220, 105418.
Popper, K. (1959). The propensity interpretation of probability. *The British Journal for the Philosophy of Science*, 10(37), 25–42.
Povinelli, D., & Dunphy-Lelii, S. (2001). Do chimpanzees seek explanations? Preliminary comparative investigations. *Canadian Journal of Experimental Psychology*, 55(2), 185.
Powell, L., & Carey, S. (2017). Executive function depletion in children and its impact on theory of mind. *Cognition*, 164, 150–162.
Powell, L., & Spelke, E. (2013). Preverbal infants expect members of social groups to act alike. *Proceedings of the National Academy of Sciences*, 110(41), E3965–E3972.
Prinz, W. (2012). *Open Minds: The Social Making of Agency and Intentionality*. Cambridge, MA: MIT Press.
Pyers, J., & Senghas, A. (2009). Language promotes false-belief understanding: Evidence from learners of a new sign language. *Psychological Science*, 20(7), 805–812.
Rakison, D., & Oakes, L. (Eds.). (2003). *Early Category and Concept Development: Making Sense of the Blooming, Buzzing Confusion*. Oxford University Press.
Rakoczy, H. (2017). In defense of a developmental dogma: Children acquire propositional attitude folk psychology around age 4. *Synthese*, 194(3), 689–707.
Rakoczy, H. (2022). Foundations of theory of mind and its development in early childhood. *Nature Reviews Psychology*, 1(4), 223–235.
Rakoczy, H., Bergfeld, D., Schwarz, I., & Fizke, E. (2015). Explicit theory of mind is even more unified than previously assumed: Belief ascription and understanding aspectuality emerge together in development. *Child Development*, 86(2), 486–502.
Rakoczy, H., Hamann, K., Warneken, F., & Tomasello, M. (2010). Bigger knows better: Young children selectively learn rule games from adults rather than from peers. *British Journal of Developmental Psychology*, 28(4), 785–798.
Rakoczy, H., & Tomasello, M. (2007). The ontogeny of social ontology: Steps to shared intentionality and status functions. In: S. L. Tsohatzidis (Ed.), *Intentional Acts and Institutional Facts: Essays on John Searle's Social Ontology* (pp. 113–137). Theory and Decision Library, vol 41. Dordrecht: Springer.
Rakoczy, H., Warneken, F., & Tomasello, M. (2008). The sources of normativity: Young children's awareness of the normative structure of games. *Developmental Psychology*, 44(3), 875–881.
Rochat, P. (2004). *The Infant's World*. Cambridge, MA: Harvard University Press.
Rochat, P., Neisser, U., & Marian, V. (1998). Are young infants sensitive to interpersonal contingency? *Infant Behavior & Development*, 21(2), 355–366.
Rochat, P., Querido, J., & Striano, T. (1999). Emerging sensitivity to the timing and structure of protoconversation in early infancy. *Developmental Psychology*, 35, 950.
Rochat, P., & Striano, T. (1999). Emerging self-exploration by 2-month-old infants. *Developmental Science*, 2(2), 206–218.

Rochat, P., & Striano, T. (2000). Perceived self in infancy. *Infant behavior and development*, 23(3-4), 513–530.1

Roebers, C. (2017). Executive function and metacognition: Towards a unifying framework of cognitive self-regulation. *Developmental Review*, 45, 31–51.

Rosati, A., & Santos, L. (2016). Spontaneous metacognition in rhesus monkeys. *Psychological Science*, 27(9), 1181–1191.

Rovee-Collier, C. (1999). The development of infant memory. *Current Directions in Psychological Science*, 8(3), 80–85.

Rubio-Fernandez, P., & Geurts, B. (2013). How to pass the false-belief task before your fourth birthday. *Psychological Science*, 24(1), 27–33.

Ruggeri, A. (2022). An introduction to ecological active learning. *Current Directions in Psychological Science*, 31(6), 471–479.

Russell, J. (1996). *Agency: Its Role in Mental Development*. Hove: Taylor & Francis.

Rüther, J., & Liszkowski, U. (2023). Ontogeny of index-finger pointing. *Journal of Child Language*, 2023, 1–17.

Saffran, J., & Kirkham, N. (2018). Infant statistical learning. *Annual Review of Psychology*, 69, 181–203.

Savage-Rumbaugh, E., Shanker, S., & Taylor, T. (1998). *Apes, Language, and the Human Mind*. Oxford University Press.

Saxe, R., & Carey, S. (2006). The perception of causality in infancy. *Acta Psychologica*, 123(1-2), 144–165.

Schelling, T. (1960). *The Strategy of Conflict*. Cambridge, MA: Harvard University Press.

Schleihauf, H., Herrmann, E., Fischer, J., & Engelmann, J. M. (2022). How children revise their beliefs in light of reasons. *Child Development*, 93, 1072–1089.

Schmidt, M., Rakoczy, H., & Tomasello, M. (2016a). Young children understand the role of agreement in establishing arbitrary norms—but unanimity is key. *Child Development*, 87, 612–626.

Schmidt, M., & Sommerville, J. (2011). Fairness expectations and altruistic sharing in 15-month-old human infants. *PLoS One*, 6(10), e23223.

Schneider, W. (1999). The development of metamemory in children. In: D. Gopher & A. Koriat (Eds.), *Attention and Performance XVII: Cognitive Regulation of Performance: Interaction of Theory and Application* (pp. 487–514). Cambridge, MA: The MIT Press.

Schraw, G., Crippen, K., & Hartley, K. (2006). Promoting self-regulation in science education: Metacognition as part of a broader perspective on learning. *Research in Science Education*, 36, 111–139.

Schulz, L., Gopnik, A., & Glymour, C. (2007). Preschool children learn about causal structure from conditional interventions. *Developmental Science*, 10(3), 322–332.

Schulze, C., & Tomasello, M. (2015). 18-month-olds comprehend indirect communicative acts. *Cognition*, 136, 91–98.

Schwier, C., Van Maanen, C., Carpenter, M., & Tomasello, M. (2006). Rational imitation in 12-month-old infants. *Infancy*, 10(3), 303–311.

Searle, J. (1995). *The Construction of Social Reality*. New York: The Free Press.

Shaw, A., Montinari, N., Piovesan, M., Olson, K., Gino, F., & Norton, M. (2014). Children develop a veil of fairness. *Journal of Experimental Psychology: General*, 143(1), 363.

Sloan, A. T., Jones, N. A., & Kelso, J. S. (2023). Meaning from movement and stillness: Signatures of coordination dynamics reveal infant agency. *Proceedings of the National Academy of Sciences*, 120, e2306732120

Slobin, D. (1985). Crosslinguistic evidence for the language-making capacity. *The Crosslinguistic Study of Language Acquisition, 2,* 1157–1249.
Sobel, D., & Kirkham, N. (2006). Blickets and babies: The development of causal reasoning in toddlers and infants. *Developmental Psychology, 42*(6), 1103.
Sodian, B., & Kristen-Antonow, S. (2015). Declarative joint attention as a foundation of theory of mind. *Developmental Psychology, 51*(9), 1190–1200.
Sommerville, J., Hildebrand, E., & Crane, C. (2008). Experience matters: The impact of doing versus watching on infants' subsequent perception of tool-use events. *Developmental Psychology, 44*(5), 1249.
Sommerville, J., Woodward, A., & Needham, A. (2005). Action experience alters 3-month-old infants' perception of others' actions. *Cognition, 96*(1), B1–B11.
Spelke, E. (1990). Principles of object perception. *Cognitive Science, 14*(1), 29–56.
Spelke, E. (2003). What makes us smart? Core knowledge and natural language. In: D. Gentner & S. Goldin-Meadow (Eds.), *Language in Mind: Advances in the Study of Language and Thought* (pp. 277–311). Cambridge, MA: MIT Press.
Spelke, E. (2022). *What Babies Know: Core Knowledge and Composition Volume 1* (Vol. 1). Oxford University Press.
Spelke, E., Breinliger, K., Macomber, J., & Jacobson, K. (1992). Origins of knowledge. *Psychological Review, 99,* 605–632.
Spelke, E., Lee, S., & Izard, V. (2010). Beyond core knowledge: Natural geometry. *Cognitive Science, 34*(5), 863–884.
Sperber, D., & Wilson, D. (1986). *Relevance: Communication and Cognition.* Cambridge, MA: Harvard University Press.
Stahl, A., & Feigenson, L. (2015). Observing the unexpected enhances infants' learning and exploration. *Science, 348*(6230), 91–94.
Stern, D. (1985). *The Interpersonal World of the Infant: A View from Psychoanalysis and Developmental Psychology.* New York: Basic Books.
Tenenbaum, J., Kemp, C., Griffiths, T., & Goodman, N. (2011). How to grow a mind: Statistics, structure, and abstraction. *Science, 331*(6022), 1279–1285.
Tennie, C., Völter, C., Vonau, V., Hanus, D., Call, J., & Tomasello, M. (2019). Chimpanzees use observed temporal directionality to learn novel causal relations. *Primates, 60*(6), 517–524.
Thomas, K., DeScioli, P., Haque, O., & Pinker, S. (2014). The psychology of coordination and common knowledge. *Journal of Personality and Social Psychology, 107*(4), 657–676.
Tomasello, M. (1992). *First Verbs: A Case Study of Early Grammatical Development.* Cambridge University Press.
Tomasello, M. (1995). Joint attention as social cognition. In: C. Moore & P. Dunham (Eds.), *Joint Attention: Its Origins and Role in Development* (pp. 103–130). Hillsdale, NJ: Lawrence Erlbaum.
Tomasello, M. (2001). Perceiving intentions and learning words in the second year of life. In: M. Bowerman & S. Levinson (Eds.), *Language Acquisition and Conceptual Development* (pp. 111–128). Cambridge University Press.
Tomasello, M. (2003). *Constructing a Language: A Usage-based Theory of Language Acquisition.* Cambridge, MA: Harvard University Press.
Tomasello, M. (2008). *Origins of Human Communication.* Cambridge, MA: MIT Press.
Tomasello, M. (2014). *A Natural History of Human Thinking.* Cambridge, MA: Harvard University Press.

Tomasello, M. (2016). *A Natural History of Human Morality*. Cambridge, MA: Harvard University Press.
Tomasello, M. (2017). What did we learn from the ape language studies? In: B. Hare & S. Yamamoto (Eds.), *Bonobos: Unique in Mind, Brain, and Behavior* (pp. 95–104) Oxford University Press.
Tomasello, M. (2018a). How children come to understand false beliefs: A shared intentionality account. *Proceedings of the National Academy of Sciences, 115*, 8491–8498.
Tomasello, M. (2018b). The normative turn in early moral development. *Special issue of Human Development, 61*, 248–263.
Tomasello, M. (2019). *Becoming human: A theory of ontogeny*. Cambridge, MA: Harvard University Press.
Tomasello, M. (2020a). The adaptive origins of uniquely human sociality. *Philosophical Transactions of the Royal Society, 375*, 20190493.
Tomasello, M. (2020b). The moral psychology of obligation. *Target Article in Behavioral and Brain Sciences, 43*, e56: 1–58.
Tomasello, M. (2022a). *The Evolution of Agency: From Lizards to Humans*. Cambridge, MA: MIT Press.
Tomasello, M. (2022b). What is it like to be a chimpanzee? *Synthese, 200*, 102.
Tomasello, M. (2022c). The coordination of attention and action in great apes and humans. *Philosophical Transactions of the Royal Society, B*. Special issue on Joint Action Coordination in Humans and Animals. (R. Heesen, Ed.): *377*, 20210093.
Tomasello, M. (2023). Social cognition and metacognition in great apes: A theory. Invited paper for special 25th anniversary issue of *Animal Cognition, 5*, 1–11.
Tomasello, M. (2024). An agency-based model of executive and metacognitive regulation. *Frontiers in Developmental Psychology, 2*, 1367381.
Tomasello, M., & Call, J. (2019). Thirty years of great ape gestures. *Special Issue of Animal Cognition, 22*, 461–469.
Tomasello, M., Call, J., & Hare, B. (1998). Five primate species follow the visual gaze of conspecifics. *Animal Behaviour, 55*(4), 1063–1069.
Tomasello, M., & Carpenter, M. (2005). Intention-reading and imitative learning. In: S. Hurley & N. Chater (Eds.), *New Perspectives on Imitation* (pp. 133–148). Oxford University Press.
Tomasello, M., & Gonzalez-Cabrera, I. (2017). The role of ontogeny in the evolution of human cooperation. *Human Nature, 28*, 274–288.
Tomasello, M., & Gonzalez-Cabrera, I. (in press). How to build a normative creature. In: C. Peacock & P. Boghossian (Eds.), *Normative Realism*. Oxford University Press.
Tomasello, M., & Haberl, K. (2003). Understanding attention: 12-and 18-month-olds know what's new for other persons. *Developmental Psychology, 39*, 906–912.
Tomasello, M., Striano, T., & Rochat, P. (1999). Do young children use objects as symbols? *British Journal of Developmental Psychology, 17*, 563–584.
Trevarthen, C. (1979). Communication and cooperation in early infancy. *Before Speech, 1*(4), 321–347.
Tronick, E. (1989). Emotions and emotional communication in Infants. *American Psychologist, 44*(2), 112–119.
Vaish, A., Carpenter, M., & Tomasello, M. (2016). The early emergence of guilt-motivated prosocial behavior. *Child Development, 87*(6), 1772–1782.
Vaish, A., Missana, M., & Tomasello, M. (2011). 3-year-old children intervene in third-party moral transgressions. *British Journal of Developmental Psychology, 29*(1), 124–130.

Vasil, J. (2023). A new look at young children's referential informativeness. *Perspectives on Psychological Science, 18*(3), 624–648.

Völter, C., & Call, J. (2014). Younger apes and human children plan their moves in a maze task. *Cognition, 130*(2), 186–203.

Völter, C., Sentís, I., & Call, J. (2016). Great apes and children infer causal relations from patterns of variation and covariation. *Cognition, 155*, 30–43.

Vygotsky, L. (1978). *Mind in Society: The Development of Higher Psychological Processes.* Cambridge, MA: Harvard University Press.

Warneken, F., Lohse, K., Melis, A., Tomasello, M. (2011). Young children share the spoils after collaboration. *Psychological Science, 22*(2), 267–273.

Warneken, F., Steinwender, J., Hamann, K., & Tomasello, M. (2014). Young children's planning in a collaborative problem-solving task. *Cognitive Development, 31*, 48–58.

Warneken, F., & Tomasello, M. (2007). Helping and cooperation at 14 months of age. *Infancy, 11*, 271–294.

Wellman, H., Cross, D., & Watson, J. (2001). Meta-analysis of theory-of-mind development: The truth about false belief. *Child Development, 72*, 655–684.

Wellman, H., & Gelman, S. (1992). Cognitive development: Foundational theories of core domains. *Annual Review of Psychology, 43*(1), 337–375.

West-Eberhard, M. (2003). *Developmental Plasticity and Evolution.* Oxford University Press.

White, P. (2006). The role of activity in visual impressions of causality. *Acta Psychologica, 123*(1–2), 166–185.

Wilkinson, A., & Huber, L. (2012). Cold-blooded cognition: Reptilian cognitive abilities. In: J. Vonk & T. Shakelford (Eds.), *The Oxford Handbook of Comparative Evolutionary Psychology* (pp. 129–143). Oxford: Oxford University Press.

Willatts, P. (1984). The Stage-IV infant's solution of problems requiring the use of supports. *Infant Behavior and Development, 7*(2), 125–134.

Willatts, P. (1999). Development of means–end behavior in young infants: Pulling a support to retrieve a distant object. *Developmental Psychology, 35*(3), 651.

Wittgenstein, L. (1953). *Philosophical Investigations.* Oxford: Blackwell.

Wobber, V., Herrmann, E., Hare, B., Wrangham, R., & Tomasello, M. (2013). Differences in the early cognitive development of children and great apes. *Developmental Psychobiology, 56*(3), 547–573.

Wolf, W., & Tomasello, M. (2020). Human children, but not great apes, become socially closer by sharing an experience in common ground. *Journal of Experimental Child Psychology, 199*, 104930.

Woo, B. M., Liu, S., & Spelke, E. S. (2023). Infants rationally infer the goals of other people's reaches in the absence of first-person experience with reaching actions. *Developmental Science, 27*(3), e13453.

Woodward, A. (1998). Infants selectively encode the goal object of an actor's reach. *Cognition, 69*(1), 1–34.

Woodward, A., Sommerville, J., Gerson, S., Henderson, A., & Buresh, J. (2009). The emergence of intention attribution in infancy. *Psychology of Learning and Motivation, 51*, 187–222.

Woolfe, T., Want, S., & Siegal, M. (2002). Signposts to development: Theory of mind in deaf children. *Child Development, 73*, 768–778.

Wyman, E., Rakoczy, H., & Tomasello, M. (2009). Normativity and context in young children's pretend play. *Cognitive Development, 24*(2), 146–155.

Wynn, K. (1990). Children's understanding of counting. *Cognition, 36*(2), 155–193.

Wynn, K. (1992). Addition and subtraction by human infants. *Nature, 358*(6389), 749–750.

Xu, F. (2019). Towards a rational constructivist theory of cognitive development. *Psychological Review, 126*(6), 841.

Xu, F., & Spelke, E. (2000). Large number discrimination in 6-month-old infants. *Cognition, 74*(1), B1–B11.

Zelazo, P. (2004). The development of conscious control in childhood. *Trends in Cognitive Sciences, 8*(1), 12–17.

Zelazo, P. (2006). The Dimensional Change Card Sort (DCCS): A method of assessing executive function in children. *Nature Protocols, 1*(1), 297–301.

Zelazo, P. (2015). Executive function: Reflection, iterative reprocessing, complexity, and the developing brain. *Developmental Review, 38*, 55–68.

Zelazo, P., Müller, U., Frye, D., Marcovitch, S., Argitis, G., Boseovski, J., . . . & Carlson, S. M. (2003). The development of executive function in early childhood. *Monographs of the Society for Research in Child Development, 68*(3), vii–137.

Index

For the benefit of digital users, indexed terms that span two pages (e.g., 52–53) may, on occasion, appear on only one of those pages.

actions 11
   action plans 14–15
   goal-directed 11, 12–13, 32–37
   intentional action 48–52
   reflexive actions 11
actualities 6, 168
adult pedagogy 130, 172–73
agency
   agentive organization 4, 12–19, 156, 163
   collective 18, 127–52
   computational model 4
   goal-directed 13–14, 21, 27–44
   individual 5–6, 13, 21
   intentional 8, 13–15, 21, 47–70
   joint 8, 17–18, 71–99
   metacognitive (rational) 13–14, 15, 21, 103–26
   shared ("we") 5, 7–8, 17, 21
Alderete 58
A-not-B error 33–34, 35, 50–51, 158–59
appearance-reality task 132
aspectuality 134
attention 12–13, 14–15
   to animate beings 29
   attentional perspective 75–76, 106, 171–72
   attention-directed contingency learning 37
   bottom-up 13, 28–29, 32–33
   joint 72, 93
   to objects and events in space 28
   top-down 32–33
   understanding others' attention 62

Baillargeon, R. 28–29
Bates, E. 55–56
Bauer, P. 49
Bayesian learning and inference 2–3, 27, 50, 113–14

behavioral decision-making 182
Behne, T. 63
beliefs
   constructive thinking and re-representation 6–7
   false beliefs 135, 137–40
   reasons for 136
   revision 112
Berkman, E. 49–50
Bermudez, J. 58–59
blame assignment 60
blicket detector machine 56–57
Bonawitz, E. 56–57
bottom-up attention 13, 28–29, 32–33
Brooks, R. 64
Bruner, J. 28
Burge, T. 30–31

Call, J. 51, 57–58, 105–6
cardinality principle 121
Carey, S. 25, 30–31, 49–50, 54, 58, 113–14, 150, 174, 177–78, 188–89
Carpenter, M. 51, 63, 80, 83
causality 35, 53–60, 160, 183
   causal force 36–37, 54, 183
   causal nets 122, 126
Cheung, P. 121
child-as-scientist metaphor 2–3, 187–89
cognitive content 160, 165
cognitive flexibility 52–53, 106
cognitive ontogeny 20–22
cognitive representations 6, 30, 71–72, 182
collaboration 72–77
collective agency 18, 127–52
collective common ground 129–30
collective intentionality 18
commitment 142

208 INDEX

common coding 39–40, 62, 64, 67–68, 161
common ground 72–73, 82–83, 129–30
communication
  cooperative/referential 78–83
  linguistic 84–94, 159, 176
computational model for agency 4
computational rationality 114–15
conceptual perspectives 75–76, 106, 130–32
conceptual representations 30–31, 150
constructive thinking 6–7, 49, 61–62, 113–16, 176, 183
control system organization 12
conversation 93
cooperative communication 78–83
cooperative coordination 7–8, 17–18
coordination games 110–11
core knowledge and concepts 30–31
Crane, T. 32
cultural common ground 129–30

Darwall, S. 142–43
decision-making
  agentive organization 12–19, 158–59
  behavioral 182
  early goal-directed actions 32–37
  either/or decisions 14–15, 33–34, 49, 158–59
  go/no-go decisions 5–6, 14, 33
  intentional action 48–52
  joint with peers 110
  metacognitive 108–10
  reflective 108
  understanding others' decision-making 65
delayed response tasks 34
DeLoache, J. S. 119–20
detour tasks 33–34, 35, 158
developmental change 169–76
developmental hypothesis 22
developmental timing 20
Devine, R. 139–40
Diamond, A. 52–53, 105
Dickinson, A. 35
Dimensional Change Card Sort 106, 117–18
discourse 93
dual representation 119–20

either/or decisions 14–15, 33–34, 49, 158–59
emotional engagement 41–42
Engelmann, J. M. 50, 144–45
evolution 5, 13–19, 39–40, 43, 71, 169, 182
executive function (tier) 5–6, 15, 51–52, 104, 139–40, 160, 178, 182
executive inhibitory control 105
executive re-representation 115
executive workspace 105–6, 160
experiential niche 13
experiential worlds 168

fairness 6–7, 143
false beliefs 135, 137–40
feeling, monitoring 108–9
Feigenson, L. 60–61
Fizke, E. 139–40
force, causal 36–37, 54, 183
framework theories 53

Gallistel, C. 121
Ganea, P. 123
gaze following 64
Gelman, R. 121
Gergely, G. 65, 66–68
Gerson, S. 38–39
Give-N Task 121
goals
  goal-directed actions 11, 12–13, 32–37
  goal-directed agency 13–14, 21, 27–44
  joint 72
  understanding others as goal-directed 38
  understanding others' goals 62
go/no-go decisions 5–6, 14, 33
Gonzalez-Cabrera, I. 141–42
Gopnik, A. 56–57, 122–23, 150
Gottlieb, G. 173
Goupil, L. 51, 108–9
grammar acquisition 89, 175
group-mindedness 128, 129–30

Haberl, K. 64
Hamann, K. 144
Hamlin, K. 33
Hauf, P. 55
Herrmann, E. 51–52
horizontal coordination 114
Hrdy, S. 41–42

Hughes, C. 139–40
Hume, D. 54
hypothesis testing 2–3, 60, 113–14, 172

iconic representations 30, 49, 159
idealized perspective 167
image schemas 30–31
imaginative representations 48
imitation 39–40, 63, 65, 66–67
individual agency 5–6, 13, 21
inference 23, 50, 57, 161–62
informative pointing 78–79, 80–81
Inhelder, B. 117
inhibition 14, 34, 49, 52–53, 105, 163
institutional facts 146
intentions/intentionality
　cognitive content 160
　collective agency 128
　collective intentionality 18
　intentional action 48–52
　intentional agency 8, 13–15, 21, 47–70
　joint intentionality 17–18, 71, 162–63
　referential intention 79–81
　social intention 81
　understanding 62–67
　understanding others' intentions 62

joint agency 8, 17–18, 71–99
joint attention 72, 93
joint decision-making 110
joint goals 72
joint intentionality 17–18, 71, 162–63

Kalnins, I. 28
Karg, K. 132, 133
Karmiloff-Smith, A. 6–7
Kim, S. 50–51
Kirkham, N. 56–57
Krachun, C. 133
Kruger, A. 149
Kuhn, T. 153, 187, 188–89

Lakatos, I. 188–89
language
　aspectuality 134
　communication 84–94, 159, 176
　developmental change 174
　false belief understanding 138–39
　normative language 146

laughing 41
launching event 35–36, 54–55, 183
Leahy, B. 49–50, 58
learning 1, 27, 37, 183
　Bayesian 2–3, 27, 50, 113–14
　developmental change 171
　hypothesis testing 2–3, 60, 113–14, 172
　iconic representations 30
　metacognitive 112
Li, L. 149
linguistic apes 89, 92–93
linguistic aspectuality 134
linguistic communication 84–94, 159, 176
linguistic construction 89
linguistic symbols 3–4, 84, 159, 174–75
locomotion 22
logical paradigms 57, 66–67
Lohmann, H. 138

Mandler, J. 30–31
matrix completion 117–18
maturation 169
means-end analysis 83
Meltzoff, A. 39–40, 63, 64, 131
memory
　metamemory 113
　working memory 48–49, 52–53, 105–6
mental coordination 7–8, 17–18, 183
metacognition 5–6, 164–65, 172–73, 182
　agency 13–14, 15, 21, 103–26
　decision-making 108–10
　inhibitory control 105
　learning 112
　metamemory 113
　re-representation 113, 178–79
　workspace 105–6
methodology, theory and 182
Millar, W. 35–36
modern theory of cognitive
　　development 2–3
Mody, S. 58
Moll, H. 65–66, 75–76, 131–32, 133
multi-perspectival representations 116–22

natural numbers 6–7, 120, 126
natural selection 11, 19–20
necessities 6, 168
normative attitudes and concepts 141–49
normative knowledge 166

normative necessities 6, 168
norms, social 145
number concept 6–7, 120, 126
Nyhout, A. 123

object classes 117, 125
objective knowledge 130–40, 165–66
objective necessities 6, 168
objective perspective 76
object permanence 33–34, 50–51, 158–59
obligation 142
O'Madagain, C. 94, 109–10, 138–39
opt-out tasks 51
ordinality 121

pedagogy 130, 172–73
perceptual representations 30–31
perspectives 75
    attentional 75–76, 106, 171–72
    conceptual 75–76, 106, 130–32
    discourse 93
    idealized (universalized) 167
    linguistic symbols 84
    multi-perspectival representations 116–22
    objective/subjective 76
physical symbols 119, 126
Piaget, J. 2, 27, 32–33, 41, 45, 48–49, 117, 121
planning 5–6, 48
Podjarny, G. 117–18
pointing 78–83
Popper, K. 181
possibilities 6, 168
practical reasoning 66–67
pretense 49, 147
priors 27, 37
profiling 87
propositional attitude construction 175–76
proto-conceptual primitives 30–31
proto-conditional 58–59, 66–67
protoconversations 41–42
proto-negation 58–59, 66–67
Proust, J. 108–9
pure coordination games 111

Quinian bootstrapping 113–14, 174

Rakoczy, H. 134, 145–46, 147, 165–66
rational (metacognitive) agency 13–14, 15, 21, 103–26

recognition respect 144–45
recursive engagement 72–73
redescription account 118–19
referential communication 78–83
reflective decision-making 108
reflexive actions 11
Reid, T. 54
relational-thematic-narrative dimension 74–75
relevance 12–13, 32
re-representation (representational redescription) 6–7, 60, 113, 176, 183
Rochat, P. 28, 30–42, 200–203
Roebers, C. 104–5
role-base schemas, grammar as 89
roles, importance of 74
rulemaking 146

Saxe, R. 54
s-causal 35
Schulz, L. 123
Searle, J. 146–47
second-order symbols 91–92
self-other equivalence 76
self-regulation 7–8
sequential guessing 49–50
shared ("we") agency 5, 7–8, 17, 21
"simulation + theory" theory 67, 161
situations 12–13
smiling 41
Sobel, D. 56–57
social and mental coordination 7–8, 17–18, 183
social cognition 170
social facts 146
social intention 81
social interaction 37–41
social norms 145
Sommerville, J. 38–39, 66
Spelke, E. 28–29, 30–31, 39, 71–72, 150, 160, 162–63, 174–75
stag hunt game 110–11
Stahl, A. 60–61
Stern, D. 41
"still face" paradigm 41–42
subjective perspective 76
symbols
    linguistic 3–4, 84, 159, 174–75
    physical 119, 126
    second-order 91–92

theory, methodology and 182
theory of cognitive development 2–3
thinking 48, 61–62, 113–16, 176, 183
three-mountains task 131
Tomasello, M. 5, 11–12, 13–14, 19–20, 39–40, 42, 64, 65–66, 75–76, 84, 87, 94, 120, 133, 138–39, 141–42, 144–45, 147, 149, 156–58, 163–64, 165–66, 169–70
tool use 48–49, 54–55, 57, 66
top-down attention 32–33
Trevarthen, C. 41
turn-taking 41
turtle task 131
two-cloth task 33–34, 50–51

uncertainty monitoring 108–9
universal grammar 175
universalized perspective 167

Vaish, A. 145–46
value-based choice 49–52
verb island constructions 90–92

vertical integration 114
violation-of-expectation 28–29, 30
visual experience 64
Völter, C. 105–6
Vygotsky, L 101, 174

Warneken, F. 144
Watson, J. 35–36
"we" (shared) agency 5, 7–8, 17, 21
weaning 22
Wellman, H. 122–23, 150
White, P. 36, 54
Wobber, V. 54–55
Woodward, A. 38–39
word learning 84
working memory 48–49, 52–53, 105–6
Wyman 147
Wynn, K. 121

Xu, F. 6–7, 30–31, 58, 113–14, 150

Zelazo, P. 53, 106, 118–19